Daniele Martins Sarmento

Evaluation of environmental radiation levels in the microPET/CT laboratory

Daniele Martins Sarmento

Evaluation of environmental radiation levels in the microPET/CT laboratory

Radioprotection

ScienciaScripts

Imprint
Any brand names and product names mentioned in this book are subject to trademark, brand or patent protection and are trademarks or registered trademarks of their respective holders. The use of brand names, product names, common names, trade names, product descriptions etc. even without a particular marking in this work is in no way to be construed to mean that such names may be regarded as unrestricted in respect of trademark and brand protection legislation and could thus be used by anyone.

Cover image: www.ingimage.com

This book is a translation from the original published under ISBN 978-3-330-76402-6.

Publisher:
Sciencia Scripts
is a trademark of
Dodo Books Indian Ocean Ltd. and OmniScriptum S.R.L publishing group

120 High Road, East Finchley, London, N2 9ED, United Kingdom
Str. Armeneasca 28/1, office 1, Chisinau MD-2012, Republic of Moldova, Europe
Managing Directors: Ieva Konstantinova, Victoria Ursu
info@omniscriptum.com

Printed at: see last page
ISBN: 978-620-8-40757-5

To my family

Detimar, Nilza, Denis and Débora

ACKNOWLEDGEMENTS

To Dr Janete Carneiro, for her dedicated guidance and help.

To M.Sc. Peterson Squair Lima, for all his patience and availability.

To Dr Gian-Maria A. A. Sordi, for always sharing his knowledge.

To Dr Demerval Leônidas Rodrigues, for being willing to help since the time of my TCC.

To my aunt Antônia Martins, for encouraging me.

To all my family and friends who always believed it was possible.

To IPEN, for the opportunity to enter postgraduate studies.

To all those who, directly or indirectly, participated and helped in the development and completion of this work.

SUMMARY

The microPET/CT system is an important piece of equipment used in diagnostic imaging research in small animals. The radiopharmaceutical most commonly used in this technology is fluorine-18-labelled fluorodeoxyglucose. The aim of this study is to carry out radiological control in the microPET/CT research laboratory at the IPEN-CNEN/SP Radiopharmacy Centre, in order to meet both national standards and international recommendations. The laboratory is classified by the facility's radioprotection team as a supervised area, in which, although the adoption of specific protection and safety measures is not mandatory, regular reassessments of working environment conditions must be carried out. In order to ensure the radiological protection of workers directly involved in handling the equipment, monitoring of the workplace and assessment of individual dose control was carried out. Initially, pre-operational monitoring was carried out, i.e. a radiometric survey of the laboratory. In addition, external radiation levels were measured in and around the laboratory facilities by placing nine $CaSO_4$:Dy thermoluminescent dosimeters (TL) in previously selected locations. Occupationally exposed individuals were assessed on a monthly basis using TL dosimeters positioned on the chest and full body measurements taken every six months. The study period was two years, starting in April 2014. To control the microPET/CT, performance tests were carried out according to the equipment's standard protocol and in accordance with the standard developed by the *Animal* PET *Standard Task Force.* This study has shown that the radiation levels in the areas (estimates of ambient dose and effective dose), as well as the equipment's shielding, are adequate in accordance with occupational exposure limits. The importance of strictly adhering to the principles of radioprotection is emphasised, since this is research with unsealed radioactive sources.

SUMMARY

CHAPTER 1

INTRODUCTION

1.1 General considerations

In the last decade there have been significant technological advances in nuclear medicine, with the aim of improving diagnostic accuracy and, consequently, prognosis. Medical imaging is the assessment of normal and abnormal tissues and organic functions of the human body by means of images. Among the various clinical diagnostic imaging techniques are Computed Tomography (CT), with 3D reconstruction, and Positron Emission Tomography (PET) [1].

A CT image is the result of computer processing of the information collected after the patient has been exposed to a series of X-rays. Tissues with different compositions absorb radiation differently, i.e. denser tissues (such as the liver), or those with heavier elements (such as calcium, present in bones), absorb more radiation than less dense tissues (such as the lung, which is full of air), thus producing an anatomical image of the organs [1, 2].

A PET image is formed by localising the annihilation process of positrons emitted by radionuclides. These radionuclides are chemical bonds between a radioactive element and a molecule capable of participating in a particular metabolic process (metabolic activity of cells and organs), which is injected into the patient [1].

In Brazil, PET technology was introduced in 1998 at the Hospital das Clínicas of the University of São Paulo Medical School. PET imaging is performed exclusively with Fluordeoxyglucose, ^{18}F-FDG [3]. This compound consists of a glucose molecule labelled with the radioisotope ^{18}F. FDG is absorbed and metabolised by a wide variety of cells.

An important characteristic of the ^{18}F-FDG radiopharmaceutical is its relatively long half-life of 110 minutes, which allows it to be transported from a cyclotron (the place where it is produced) to the Nuclear Medicine Service (NMS), where it will be used, in a maximum of 2-3 hours [4].

In oncology, the PET image is formed by the increased uptake of glucose by cancer cells, since they metabolise glucose at higher rates than cells in normal tissue. In this way, a procedure using FDG can identify primary or metastatic cancer before structural evidence of the disease is present [5].

Similar to the activity of cancer cells, other inflammatory processes, such as metabolic activity in the brain, heart and other organs, can pick up glucose independently of the existence of tumours, which makes it possible to use ^{18}F-FDG in PET scans for non-oncological purposes [4].

PET and CT components can be operated separately or in combination, mounted on a single gantry (mechanical movement system, rotational support). PET-CT technology combines the techniques of Nuclear Medicine (PET) with Radiology (CT), giving it an advantage over other imaging diagnostic modalities, since it has the ability to assess the metabolism of lesions, demonstrating the presence of functional alterations even before the anatomy is affected, which allows for an early and more accurate diagnosis [5].

The success of PET scanners in humans naturally led to interest from the pharmaceutical industry and biomedical companies to start developing PET scanners for small animals in the 1990s, the microPET scanners, in order to continue the research. The main factors driving their development were the inadequacy of existing scanners for studies on mice, rats, etc., as well as the legal issues surrounding the use of human devices on small animals. The price of the scanner was also an important factor to consider, as the difference in the physical dimensions of a microPET, compared to a PET for human application, would be much smaller [6].

These systems also have an advantage in research with small animals, with better spatial resolution, making it possible to carry out pre-clinical tests with drugs (new medicines) intended for the treatment of tumours in humans. The parallel development of microPET systems is also clearly advantageous for the development of PET scanners for use in humans, making technology transfer feasible [6].

The high-energy tracers used for PET/CT studies essentially require adequate shielding to ensure that working conditions are acceptably safe and satisfactory for occupationally exposed individuals (OEIs) when carrying out their tasks. The highly energetic gamma rays emitted by ^{18}F (511 keV) can entail a higher whole-body dose for workers than for those using tracers labelled with ^{99m}Tc or ^{123}I in conventional nuclear medicine [7].

This research aims to use environmental and individual monitoring techniques to assess the level of radiation in the environment, as well as estimating the doses of the IOEs involved in microPET/CT research (radiological conditions in the workplace).

The environmental monitoring programme will ensure that working conditions are acceptably safe and satisfactory for occupationally exposed individuals and that the dose levels established by the regulatory authority in Brazil, the National Nuclear Energy Commission - CNEN, for both free and controlled areas, are not exceeded [8].

1.2 Purpose and objectives

The purpose of this work is to carry out radiological control on the physical premises of the Research & Development laboratory at IPEN's Radiopharmacy Centre (RC), which currently has an Albira microPET/SPECT/CT system. This study involves the control of microPET/CT, used for research purposes on small animals.

In order to achieve the aim of this study, it will be necessary to establish the following objectives:

1. Carry out pre-operational monitoring in the workplace, i.e. assess the radiation levels in the areas in contact with the laboratory where the microPET/SPECT/CT is installed, using portable equipment;

2. Carry out area monitoring of the laboratories, identifying the places most likely to be exposed to external radiation, and place the detectors, thermoluminescent dosimeters (TLD), in the environment (operational test);

3. Evaluate occupational exposure both in the area and individually when using the microPET/CT system;

4. To carry out performance tests of the microPET system in accordance with the NEMA NU 4-2008 standard [9], for small animals, in order to analyse parameters such as spatial resolution, sensitivity and scattering fraction, using a standard simulator and a point source.

5. Compare the results of the performance tests obtained with the standard protocol pre-established by the equipment manufacturer.

CHAPTER 2

LITERATURE REVIEW

2.1 Positron emission tomography, PET

Positron emission tomography is a study in which the coincidence detection of two photons with an energy of 511 keV, released from the annihilation process between positrons emitted by the radiopharmaceutical and electrons in the medium, is used to verify the distribution and kinetics of the positron-emitting radiopharmaceutical in vivo. [4, 10]

The PET equipment's detection system consists of a ring with detectors (scintillator materials[I]) distributed along it, associated with electronic circuits that identify the photons coming from each annihilation. Each annihilation is recorded as a single event, occurring at a point on the line joining the two detection sites, the Line of Response (LOR).

This system is called a coincidence detection system (Figure 2.1). Secondary photons can also be distinguished without the need for lead collimators, which increases sensitivity; the collimation used is electronic, where pairs of photons that are detected in different positions within a very short time interval characterise a coincidence and constitute an event [4].

[I]Scintillator materials: usually crystals, they transform the kinetic energy of the incident radiation into light that can be detected by photomultipliers [11].

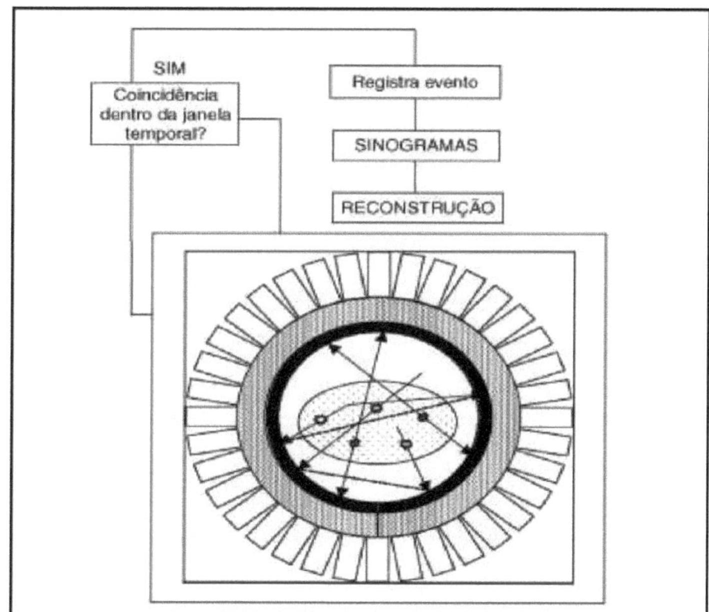

Figure 2.1 Coincidence detection scheme (photon pairs) in PET systems [3].

Currently, modern PET systems are hybrid equipment, coupled with CT and/or MRI scans and are made up of more than 20,000 detection elements arranged in adjacent rings, which record coincidence events within intervals of the order of nanoseconds [3, 4].

The detector array is mounted on a gantry and makes a complete circle or part of a circle around the patient, allowing data to be acquired from several simultaneous projection angles and reducing artefacts caused by patient movement. The addition of a coincidence circuit to a conventional two-head gamma camera also makes it possible to use it to acquire positron emitter images, as long as the thickness of the crystal and the system allow for this type of modification. Among the existing models, there may be differences in the type of crystal (the most commonly used crystals are Lutetium Orthosilicate (LSO) or Sodium Iodide (NaI) crystals), the number of detector rings and whether or not they are coupled to a CT or MRI scanner [3].

PET is currently the most advanced and sophisticated procedure in the field of nuclear medicine. It is a powerful metabolic imaging technique that uses positron-emitting radionuclides as metabolically active tracers, i.e. a biological molecule that carries a positron emitter such as ^{11}C, ^{13}N, ^{15}O or ^{18}F [4, 10].

Computed tomography, CT, uses X-rays with computer technology, shows the anatomy of organs and makes it possible to take precise images of the location, size and shape of tumours/lesions in an organ.

In Brazil, practically all the installed PET-Scans correspond to PET/CT equipment.

As a technical matter, every PET image must have a so-called "attenuation correction". Without this, the images lose quality and cannot be quantified. In PET/CT equipment, attenuation correction is carried out using CT images [12].

1.2 PET/CT image fusion and its advantages

The device performs both tests to assess cellular metabolism and whole-body anatomy, enabling more accurate diagnoses, the detection of diseases at an early stage, radiotherapy planning, monitoring and the choice of a more effective treatment, depending on each case.

The fusion of PET and CT images allows for a better diagnosis due to the integration and visualisation of the scans. While PET shows images of cellular metabolic activity (functional information), CT shows the anatomy of the organs, making it possible to locate the region, size and shape of the tumour. In addition, it is not necessary to move and reposition the patient in each of the systems separately. The use of CT data to correct for the attenuation of PET data avoids long PET acquisitions, which used to be done in relation to external radioactive sources, and consequently reduces the total PET/CT acquisition time. The PET/CT system also allows for a more complete clinical report, as it contains a fusion of functional and anatomical images [4].

2.3 Main applications of PET and PET/CT

The PET technique has basically been used in three important areas of clinical diagnosis: oncology [85%], followed by neurology/psychiatry [10%] and cardiology [5%]. The greatest application is in oncology, providing detection, localisation and staging of tumours; in neurology and psychiatry, it makes it possible to assess dementia, seizures, movement disorders and diagnose diseases such as Alzheimer's and in cardiology, where it is mainly used in problems related to the coronary arteries [13].

The use of PET/CT systems has grown in recent years in the medical community. The fusion of functional and anatomical images carried out in PET and CT, respectively, allows for the development of a better diagnosis in clinical oncology, as well as better patient care and benefit [1].

2.4 Albira imaging equipment: MicroPET/SPECT/CT

The Albira imaging equipment, illustrated in Figure 2.2, combines the imaging techniques of Positron Emission Tomography (PET), Single Photon Emission Tomography (SPECT) and Computed Tomography (CT), for use with small animals across a wide range of (pre-clinical) research fields. The modular system design allows the choice of one or other combination of these modalities [14, 15].

Albira Imaging Equipment has four units:

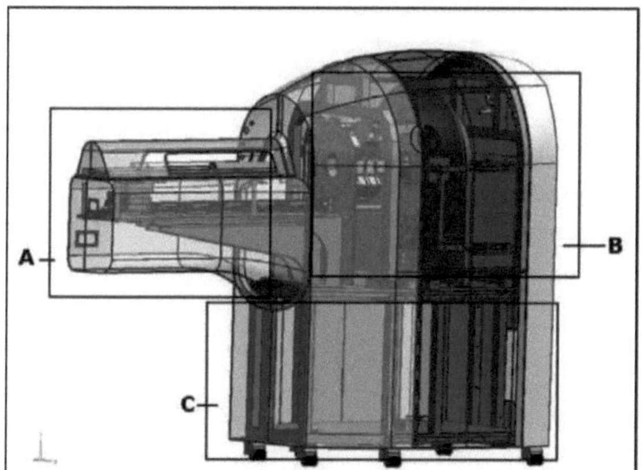

Figure 2.2 Albira imaging equipment, units A, (access area) B (image acquisition system) and C (electronic sensors) [15].

A) The animal preparation unit (access area or stretcher) contains the translational movement system that allows access to the animal for routine procedures.

B) The image acquisition system on which the PET, CT and SPECT modules can be installed. The equipment present in the Microbiology laboratory of the Radiopharmacy Centre currently has the three modules, all with installation dates: PET on 25/02/2013, CT on 29/11/2013 and SPECT on 08/08/2014. The equipment is monitored by sensors with a redundant alarm system to restrict access to authorised service personnel.

C) The lower unit of the system contains the sensor subsystems and associated electronics for image acquisition and control.

D) The control unit consists of two external computers, one of which is used by the acquisition and processing system and the other for storing the data produced (images).

Figure 2.3 shows an image acquisition diagram similar to the one available at IPEN.

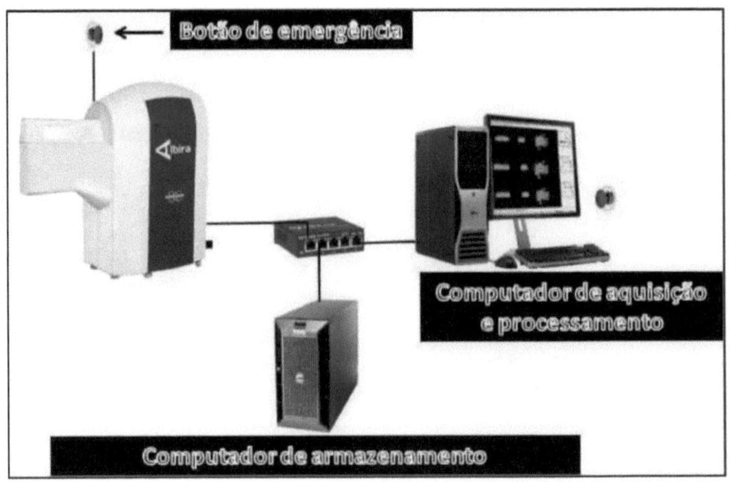

Figura 2.3 Pre-clinical image acquisition system [15].

2.3.1 PET Albira module

The Albira PET imaging module is housed in the initial region of the image acquisition system. The PET is a modular device that can be configured with 1, 2 or 3 rings (8 PET detectors per ring - "installed with 2 rings", Fig.2.4.a.) The system's performance (resolution, sensitivity and FOV - field of view) improves with the addition of each ring. Each PET sensor consists of a continuous crystal, a photomultiplier (PSMT), and a signal collection network, as shown in Figure 2.4 (a) and (b) [14, 15].

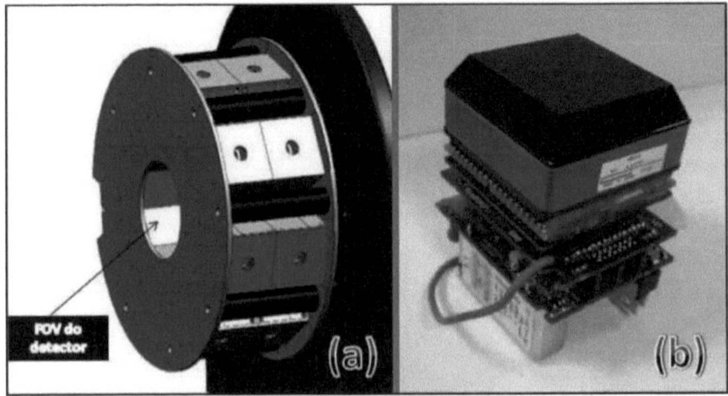

Figure 2.4 (a) Albira PET imaging module; (b) PET sensor [14, 15].

1.1.2 Albira SPECT module

The SPECT imaging module is housed in the central region of the image acquisition system

(Figure 2.5). The central components of the camera array are: the collimators (single pinhole or multi-pinhole), a continuous scintillation crystal and a photomultiplier (PSMT). SPECT plates are designed for high portability and low electronic noise. Therefore, the Albira SPECT module as a whole has low electronic noise, which is necessary to obtain high-resolution images [15].

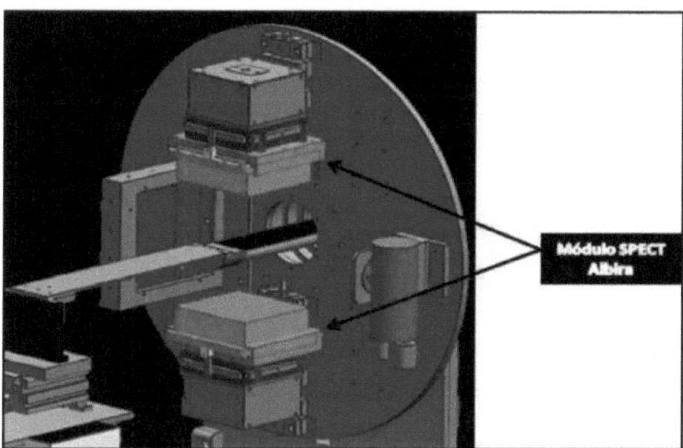

Figure 2.5 Albira SPECT module [15].

2.4.3 CT Albira module

The CT imaging module is housed in the imaging system (Figure 2.6) so that it is in a parallel plane with the two PET and SPECT modules, as it provides an anatomical image that acts as a useful reference for functional imaging. The CT uses a X-ray tube for industrial applications with a nominal focal point of 35 μm, suitable for imaging small animals. The anode is made of tungsten. The voltage for the Albira CT module's X-ray tube can be set to 35 or 45 kV with 400 or 800 μA. Normally the X-ray tube is designed to work between 10 and 50 kV with a maximum current of 1 mA. X-ray energy adjustments can be made to minimise the X-ray dose or maximise image quality when necessary. The detector (Flat Panel) is built in a matrix format allowing for high-resolution subunits (50 μm) over large areas. The sensor is made up of a matrix of 2400 x 2400 pixels. This translates into a 2D detection area of 120 mm x 120 mm, making it possible to visualise the entire image of an animal (mouse) in a single rotation [15, 16].

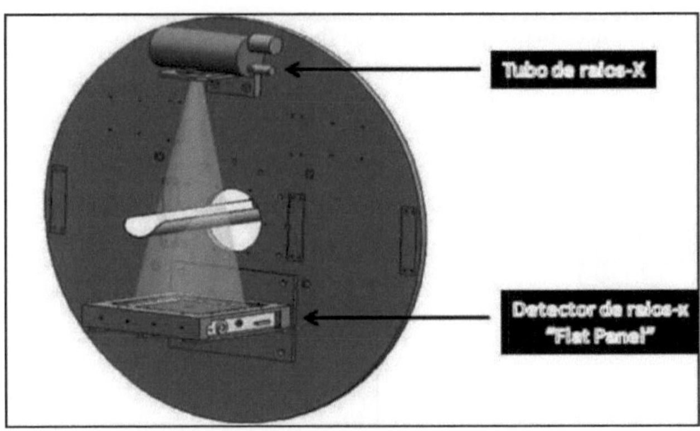

Figure 2.6 Albira CT imaging module [15].

The system is assembled with the image acquisition parts and two computers. Operational control of the Albira imaging system is carried out via an Albira Suite software interface [15, 16].

1.5 Study of the physical parameters of microPET

Although the performance of a PET scanner depends not only on the detectors, but on the design of the scanner as a whole, it is important to check for variations in the physical parameters supplied by the manufacturer. Therefore, some tests, including resolution, sensitivity, uniformity and field of view (FOV), are essential to verify compliance with the system's main operating characteristics. The use of a standard evaluation simulator can indicate whether there are variations in the performance and control of the equipment [17].

In Brazil, the standards for human PET/CT can be followed by carrying out quality control tests for PET, PET/CT and CT (Quality Assurance for PET and PET/CT Systems [18], CNEN NN 3.05 [19] and Anvisa (1998) [20]; these standards describe the limits for occupational individuals and the functional parameters of the equipment.

With regard to pre-clinical equipment, it is important to evaluate and standardise the tests carried out in the country, since equipment for small animals has different standards to human PET scanners.

In this context, it is recommended to use the publication of the National Electrical Manufacturers Association (NEMA). This American national standard provides a forum for the development of technical standards of interest to the industry and its users [21].

In the case of microPETs, it is recommended to use NEMA publication NU 4-2008 [9], which proposes a standardised methodology for evaluating the performance of microPET systems designed for small animal imaging [9, 21].

The standard establishes a baseline of system performance under normal imaging conditions, independent of camera design, and applies to a wide variety of camera models and geometries. It represents a subset of measures that characterise the performance of PETs with

specific objectives, which are found in small animal imaging laboratories. This subset is considered common to all existing CT scanners at the time of publication [9].

2.6 Radioprotection - Occupational exposure resulting from PET/CT practices

As previously mentioned, ^{18}F is the radionuclide used in Brazil and its administration in PET studies/procedures is carried out using the radiopharmaceutical fluordeoxyglucose, FDG. Due to the high energy of the photons produced and the increase in the number of PET examinations in Brazil, it is necessary to optimise the techniques for using ^{18}F-FDG to ensure compliance with the radiological protection guidelines set out in the CNEN-NN-3.01 [8] and Anvisa 1998 [20] standards. Therefore, one of the aims of this work is to assess the occupational exposure (external irradiation and internal contamination) of the Occupationally Exposed Individuals (OEIs) involved in the research, identifying the stages of the process where the highest levels of exposure may occur.

According to the study by KUBO [4] and other studies found in the literature [22, 23] doses received by Occupationally Exposed Individuals (OEIs) in Nuclear Medicine Services can be high, especially in the hands, which justifies a constant concern to optimise radioprotection and realistically monitor individual doses. External exposures can be minimised, for example by training IOEs and additional shielding in order to reduce doses during procedures. The tasks in which IOEs receive the highest levels of radiation exposure include preparing the radiopharmaceutical; checking the activity of the radiopharmaceutical to be injected; administering radiopharmaceuticals to patients; carrying out tasks near patients after they have been injected; monitoring the patient before, during and after the examination; as well as quality control of equipment using sealed or unsealed sources.

Data from the UNSCEAR 2008 Report [24] show that the occupational exposure of technicians carrying out PET procedures can be quite high, with values of 3, 10 and 12 mSv. These dose values are higher than the values found, on average, of 3 and 2 mSv for technicians carrying out other procedures in conventional nuclear medicine. Although the exposure caused by ^{18}F-FDG emissions is higher than the exposure caused by other radiopharmaceuticals (^{67}Ga, ^{208}Tl, etc.) used in conventional nuclear medicine procedures, it does not represent a greater risk for IOEs. [24].

The UNSCEAR 2008 publication [24] reports that the occupational doses received in CT procedures are very small and the technique does not represent a significant source of occupational exposure. This is due to the high collimation of the primary X-ray beam and consequently low levels of scattered radiation. In all CT units, leakage radiation is practically zero. The control room, where the technician sits, of a CT facility does not represent a significant source of radiation [24].

CHAPTER 3

THEORETICAL FOUNDATIONS

3.1 Interaction of radiation with matter

3.1.1 General considerations

Ionising radiation is electromagnetic waves (X-ray or gamma-ray photons) or particles; electrons, protons and neutrons capable of exciting, ionising and/or activating atoms of the matter with which they interact.

Radiation can be produced by adjustment processes that take place within the nucleus or in the electronic layers, by the interaction of the nucleus or atom with other radiation or particles. Beta radiation and gamma radiation (adjustment in the nucleus) and characteristic X-rays (adjustment in the electronic structure) are good examples [11].

In general, when electromagnetic radiation interacts with matter, three types of phenomena can occur: excitation, ionisation and activation [4].

a) Excitation: In this case, the ionising radiation does not transfer enough energy to knock electrons out of the atom. The electrons are displaced from their equilibrium orbitals to a more energetic one and, when they return, they emit the excess energy in the form of radiation.

b) Ionisation: When the radiation has enough energy to knock the electron out of the atom, i.e. cause ionisation.

c) Activation: This occurs when the incident radiation interacts with the nucleus of the atom, forming radionuclides. When the atom seeks stability, it emits excess energy in the form of radiation.

Ionising radiation can also be classified as directly ionising and indirectly ionising [4]:

Directly ionising radiation: fast charged particles (e.g. B^+, R>- and a), which release their energy directly into matter through many small Coulombian interactions along their path.

Indirectly ionising radiation: X-ray photons, *Y-ray* photons and neutrons, i.e. uncharged particles that first transfer their energy to charged particles in matter, through which they pass making relatively few interactions. The charged particles then transfer their energy to matter as in the previous section.

Radionuclides undergo radioactive decay by losing discrete amounts of energy in the form of one or more atomic particles or photons. The main emitted ionising radiation to be considered are X-rays, Y-RAYS, ß-radiation$^+$, ß$^-$ and α-particles and neutrons [4].

X-rays

X-rays are high-energy photons that come from electronic transitions in atoms. X-rays are known to be electromagnetic radiation, with wavelengths longer than ultraviolet radiation.

X-rays can be classified in two ways, characteristic X-rays (emitted by charged particles when they change atomic energy level) and brake or bremsstrahlung X-rays (emitted due to Coulombian interaction with nuclei of high atomic number or with other electrons in the electrosphere) [1, 4].

γ-rays **(gamma)**

It has the same electromagnetic nature as radiation, differing only in its origin, since gamma rays are of nuclear origin, while X-rays are produced in the electrosphere.

When a nucleus decays by emitting alpha or beta radiation; when the residual nucleus has its nucleons out of equilibrium configuration, they are allocated to excited states. In order to reach the ground state, they emit excess energy in the form of electromagnetic radiation, called γ-rays, as illustrated in Figure 3.1 [4, 11].

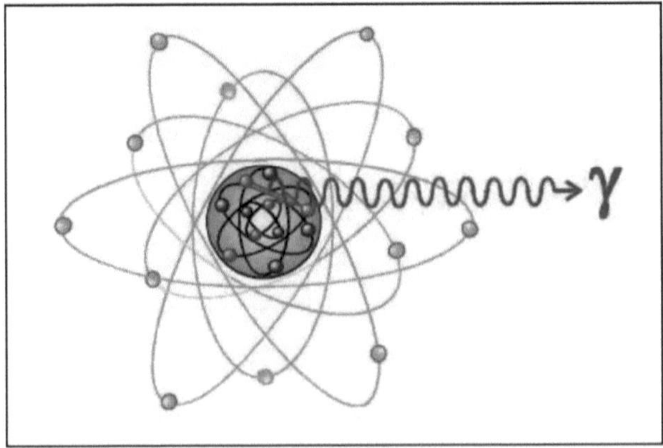

Figure 3.1 Nucleus reaches ground state, emits excess energy in the form of γ radiation [11].

Radiation ß⁺ , ß⁽⁻

Beta radiation is the term used to describe positively or negatively charged electrons (neutrons and positrons) of nuclear origin. Its emission is a common process in small or intermediate mass nuclei, which have an excess of neutrons or protons in relation to the corresponding stable structure. Figure 3.2 illustrates the beta decay process [4, 11].

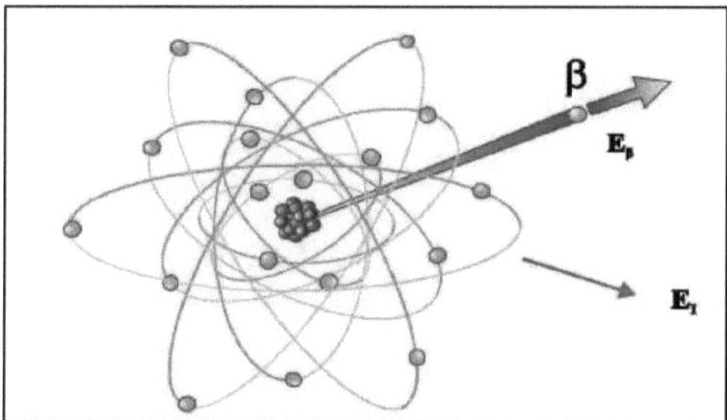

Figure 3.2 Emission of β radiation[11].

Emission $ß^-$

When a nucleus has an excess of neutrons inside it, the compensation mechanism takes place through the transformation of a neutron into a proton, releasing an anti-neutrino (ʊ̈) and an electron (called the $ß^-$ particle), emitted in the decay process. In order to conserve energy during the decay process, the anti-neutrino (a particle with a mass of approximately zero and no charge) is emitted in such a way as to share with the emitted electron the energy released by the nucleus [4, 11].The transformation of the neutron into a proton by the $ß^-$ emission process can be represented by the equation:

$$ {}^{0}_{1}n \rightarrow {}^{+}_{1}p + {}^{-}_{0}e + \bar{v} $$

Issue B^+

It occurs in nuclei with an excess of protons. In this case, the proton transforms into a neutron by emitting a ß particle⁺ (called a positron) and a neutrino (v), with a function similar to the anti-neutrino, explained in the previous section [4, 11].The transformation of the proton into a neutron by the R+ emission process can be represented by the equation:

$$ {}_{1}^{+}p \rightarrow {}_{1}^{0}n + {}_{0}^{+}e + \nu $$

After being ejected from the nucleus, the positron loses its kinetic energy in collisions with surrounding atoms of matter and comes to rest. This happens just a few millimetres from where it originated. The positron combines with the electron in an annihilation reaction, where their masses are converted into two photons of 511 keV each, leaving the annihilation site in approximately opposite directions (180°±0.5°) [4].

Particle a

It occurs in nuclei with a high number of protons and neutrons. When this happens, the nucleus can become unstable because the electrical repulsion between the protons can overcome the attractive nuclear force. In this case, it is possible to emit the a-particle, which is made up of 2 protons and 2 neutrons (essentially the nucleus of the ^{4}He atom), with a large amount of energy [4].

In this type of decay, the nuclear modifications are described as in the following equation:

$$ {}_{Z}^{A}X \rightarrow {}_{Z-2}^{A-4}Y + \alpha + energia $$

The a-particles also have great ionisation power, which is the main way in which they lose energy before being absorbed [4, 11].

Neutrons

The neutron has a large mass and does not interact with matter through the Coulombian force, which predominates in the processes of energy transfer from radiation to matter. For this reason, is very penetrating and, unlike gamma radiation, secondary radiation is often recoil nuclei, especially for hydrogenated materials with high ionisation power. In addition to recoil nuclei, there are the products of (n, a) type nuclear reactions, which are highly ionising. Unlike other ionising radiation, the neutron can easily interact with the atomic nucleus and sometimes activate it [4, 11].

3.1.2 Main interactions of electromagnetic radiation

Photons interact indirectly with matter, transferring all or part of their energy to electrons, which in turn cause ionisation. The presence of a photon beam can only be perceived when its energy is released in an absorbing medium through certain interactions. As a result, radiation passing through a detector can only be detected if there is a loss of energy from that radiation to the medium. Similarly, radiation that passes through human tissue without transferring energy to it does not release a dose into the medium [4].

In nuclear medicine, the main ways in which photons interact with matter are: photoelectric effect, Compton scattering and pair production. The probability of these effects occurring depends on both the photon's energy (E) and the atomic number (Z) of the absorbing medium. Figure 3.3 shows the Z and E regions in which each interaction predominates [4].

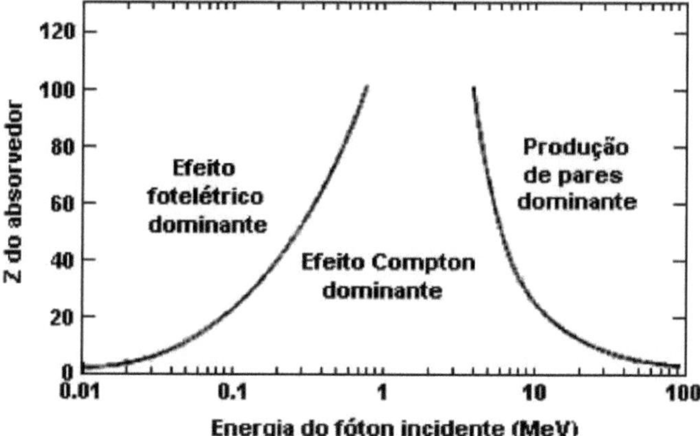

Figure 3.3 Atomic number and energy values at which each interaction predominates [4].

Photoelectric effect

The photoelectric effect corresponds to the interaction of photons with electrons in the atom's highest energy orbits (innermost orbits). The photon transfers all its energy to the electron. Part of this energy is used to overcome the electron's binding energy and the remaining energy is transferred to the electron itself in the form of kinetic energy. The probability of this process occurring decreases with increasing photon energy, but increases with increasing atomic number of the material. The photoelectric effect is dominant in human tissues at energies below approximately 100 keV and plays a very important role in X-ray imaging [1].

Compton effect

The Compton effect corresponds to the interaction of the photon with the electron in one of the lower energy orbits (outermost orbit). As the binding energy of these electrons is very small, the electron is considered to be free. This effect predominates in human tissues at energies between 100 keV and 2 MeV, approximately. The potential energy of the electron is very small compared to the energy of the photon and can be considered negligible for the calculation. After the interaction, the photon is deflected and the electron is ejected from the atom. The loss of energy by the photon is divided between the electron's equilibrium energy and the kinetic energy it acquires. The energy transferred does not depend on the properties of the material or the electron density. When the photon is deflected from its path, it can interact with matter again in the form of the Compton effect or the photoelectric effect, or it can simply not interact with matter again [1].

Pair production

Pair production is an absorption process in which a photon disappears and gives rise to a pair, both with the mass of an electron but with opposite charges, i.e. an electron-positron pair (e^{-}-e^{+}), which are ejected from the site of the interaction. This phenomenon occurs when a photon with a minimum energy of 1.022 MeV (twice the energy of the resting electron mass, $m0c^2$) interacts with an intense electric field from a charged particle, usually an atomic nucleus [4].

As the presence of a high electromagnetic field is necessary for the production of pairs, the probability of this effect occurring increases the higher the charge involved, i.e. the higher the atomic number of the material [4].

The positron created has a short half-life because, after losing its kinetic energy, it collides with an electron and both annihilate, generating two photons with equal energies, with the highest probability of occurrence of 0.511 MeV, emitted in diametrically opposite directions [4].

3.2 Main quantities and units of radiological protection

The quantities used for dose limits are called dose limitation quantities. Although they can be calculated, these quantities are not measurable. However, they can be estimated from basic dosimetry quantities, such as absorbed dose, kerma or fluence, as well as incorporated activity or activity present in the environment [25].

3.2.1 Activity, A

The activity of a radioactive material is the amount of radionuclide in a given energy state at an instant of time. This quantity is defined by:

$$\boldsymbol{A} = \frac{\boldsymbol{dN}}{\boldsymbol{dt}}$$

Where dN is the expected value of the number of spontaneous nuclear transitions of that energy state in the time interval dt [8, 26] and the unit in the International System, SI, is called Becquerel (Bq).

3.2.2 Exposure, X

The exposure quantity is defined as the quotient of dQ by dm, where dQ is the value of the total charge of ions of a given sign produced in the air when all the electrons (negative and positive) released by the photons in a volume of air of mass dm are completely stopped in the air, according to the equation below [11].

$$X = \frac{dQ}{dm}$$

This quantity is only valid for X and gamma radiation and its SI unit is C/Kg [26].

3.2.3 Kerma, K

The quantity Kerma (kinectic energy released per unit of mass) is defined as the quotient dE_{tr} by dm, where dE_{tr} is the sum of all the initial kinetic energies of all the charged particles released by neutral particles or photons incident on a material of mass dm [26].

$$K = \frac{dE_{tr}}{dm}$$

The corresponding unit, the Joule/kilogram (J/Kg), also goes by the special name of Gray (Gy), where 1 Gy = 1 J/kg^{-1}.

As kerma includes the energy received by charged particles, usually ionisation electrons, these can dissipate it in successive collisions with other electrons, or in the production of braking radiation (bremsstrahlung),

$$K = Kc + Kr$$

Where, Kc is the collision kerma, when the energy is dissipated locally, by ionisations and/or excitations, Kr is the radiation kerma, when the energy is dissipated far from the site, by means of X-rays.

3.2.4 Creep, Φ

Fluence is the ratio of dN/da, where dN is the number of particles incident on a sphere of cross-sectional area da, measured in units of m^{-2} [26].

$$\Phi = \frac{dN}{da}$$

The number of particles N can correspond to particles emitted, transferred or received. This quantity is often used to measure neutrons.

3.2.5 Absorbed dose, D

The absorbed dose quantity, D, was proposed to overcome the limitations of the exposure quantity; it is valid for all types of ionising radiation and for any type of absorbing material. It is defined as the ratio of $d\bar{\varepsilon}$ to dm where $d\bar{\varepsilon}$ is the average amount of energy deposited in a given volume of matter, of mass dm , by ionising radiation, according to the equation below.

$$D = \frac{d\bar{\varepsilon}}{dm}$$

The current unit of absorbed dose is the Gray (Gy), which is equivalent to 1 J/k [8, 26, 27].

3.2.6 Equivalent dose, H_T

The equivalent dose quantity, H_T, was defined to provide information on the biological damage caused to humans by each type of radiation. It takes into account factors such as the type of ionising radiation, the energy and the distribution of the radiation in the tissue.

The equivalent dose is expressed as:

$$\boldsymbol{H_T = D_T w_R}$$

Where D_T is the average absorbed dose in the organ or tissue and w_R is the radiation weighting factor [8, 26, 27].

The w_R values were defined on the basis of the relative biological effectiveness (RBE) of the different types of radiation.

The unit of magnitude of the equivalent dose in the SI is the J/kg, called Sievert (Sv).

3.2.7 Dose equivalent

Operational quantities for external monitoring can be defined with metrological characteristics and also take into account the different damage efficiencies for different types and energy of radiation. These are the personal dose equivalent $H_p(d)$ and the ambient dose equivalent H*(d). These quantities use the radiation quality factors Q as a weighting factor instead of the radiation weighting factors w_R. The radiation quality factors are given as a function of the linear transfer of unrestricted energy (also called unrestricted braking power) and are established in CNEN Regulatory Position 3.01/002:2011 [25].

Personal dose equivalent - $H_p(d)$ - is an operational quantity for individual external monitoring, and is the product of the absorbed dose at a point, at depth d of the human body, by the radiation quality factor at that point [25].

Ambient Dose Equivalent - H*(d) - is an operational quantity for monitoring in the work environment area, and is the product of the absorbed dose at a point, by the radiation quality factor, corresponding to what would be produced in an equivalent sphere of tissue 30 cm in diameter, at depth d [25].

3.2.8 Effective dose, E

In order to limit the risk of stochastic effects, the concept of effective dose was introduced. This quantity is based on the principle that for a certain level of protection, the risk should be the same if the whole body is irradiated uniformly, or if the irradiation is localised to a particular organ.

The effective dose is given by the sum of the weighted equivalent doses in the various organs and tissues and is defined by the following equation:

$$E = \sum_{T} w_T H_T$$

Where is the weighting factor of the organ or tissue, which takes into account the risk of stochastic effect, and HT is the equivalent dose in the tissue or organ [8, 26, 27].

The unit of the equivalent dose quantity in the SI is the J/kg, called Sievert (Sv).

1.3 Basic principles of radiological protection [8]

The main purpose of radiological protection is to protect individuals from the harmful effects of ionising radiation, allowing them to work safely in activities using radiation. The philosophy of radiological protection adopts the following principles:

Principle of justification: No practice or source associated with that practice should be authorised unless the practice produces benefits for the individuals exposed or for society sufficient to outweigh the corresponding detriment, taking into account social and economic factors, as well as other relevant factors.

Principle of optimisation: In relation to exposures caused by a given source associated with a practice, radiological protection must be optimised so that the magnitude of individual doses, the number of people exposed and the probability of exposures occurring remain "as small as reasonably achievable", according to the ALARA (As Low As *Reasonably Achievable)* principle, taking into account economic and social factors. In this optimisation process, it should be noted that the doses to individuals from exposure to the source should be subject to the dose restrictions related to that source.

Principle of individual dose limitation: The normal exposure of individuals should be restricted in such a way that neither the effective dose nor the equivalent dose in the organs or tissues of interest, both caused by the possible combination of exposures originating from authorised practices, exceed the annual dose limits established and presented in Table 3.1.

Table 3.1 Annual dose limits [8]

Greatness	Organ	Occupationally exposed individual	Individual from the public
Effective dose	Whole body	20 mSv [b]	1 mSv [c]
Equivalent dose	Crystal	20 mSv	15 mSv
	Skin [d]	500 mSv	50 mSv
	Hands and feet	500 mSv	---

[a] For the purposes of administrative control carried out by CNEN, the term annual dose should be considered as the dose in the calendar year, i.e. in the period from January to December of each year.
[b] Arithmetic mean over 5 consecutive years, provided it does not exceed 50 mSv in any year.
[c] In special circumstances, CNEN may authorise an effective dose value of up to 5 mSv in one year, provided that the average effective dose over a period of 5 consecutive years does not exceed 1 mSv per year.
[d] Average value in $1cm^2$ of area, in the most irradiated region.

The effective dose values apply to the sum of the effective doses caused by external exposures and the effective committed doses (integrated over 50 years for adults and up to the age of 70 for children) caused by incorporations in the same year.

3.4 Dose restriction, occupational reference levels and area classification [28]

3.4.1 Dose restriction

In order to guarantee an adequate level of individual protection for each IOE, an effective dose restriction value is established as a limiting condition for the radiological protection optimisation process, taking into account the uncertainties associated with it, for any source or facility under regulatory control.

3.4.2 Registration and Investigation Levels

The recording level for monthly individual IOE monitoring is 0.20 mSv for effective dose; all doses greater than or equal to 0.20 mSv must be recorded, although recording doses below this level is also possible.

The investigation level for individual IOE monitoring should be, for effective dose, 6 mSv per year or 1 mSv in any month. For equivalent dose, the investigation level for skin, hands and feet is 150 mSv per year or 20 mSv in any month. For the crystalline lens, the investigation level is 6 mSv per year or 1 mSv in any month.

3.4.3 Area classification

The area classification system is proposed to help control occupational exposures. It considers the designation of workplaces into two types of areas: controlled areas and supervised areas.

Controlled area - an area subject to special protection and safety rules in order to control normal exposures, prevent the spread of radioactive contamination and prevent or limit the extent of potential exposures.

Supervised area - an area where occupational exposure conditions are kept under supervision, even if specific protection and safety measures are not normally required.

When the possibility of contamination by radioactive materials is remote, areas can sometimes be defined in terms of the dose rate around them. The use of mobile sources requires some flexibility in defining these areas.

Areas must be classified whenever occupational exposure is expected and clearly defined in the Radiological Protection Plan. This classification should always be reviewed if necessary, depending on how

of operation or any modification that could alter the conditions of normal or potential exposure. Outside areas designated as controlled or supervised, the dose rate and the risk of contamination by radioactive materials must be small enough to ensure that, under normal conditions, the level of protection for those working on site is comparable to the level of protection required for public exposures, i.e. 1 mSv/year. These areas are called free areas from the point of view of occupational radiological protection.

1.5 Ionising radiation monitoring programme for workers

Monitoring is an ongoing process of a preventive or confirmatory nature. Monitoring techniques make it possible both to alert workers and members of the public to the presence of radiation, in order to avoid a dose that exceeds the limits set by national standards, and to assess the doses already received, which could not be avoided.

Preventive monitoring is proportional to the dose estimated to affect the occupationally exposed individual, IOE, and therefore when the sum of the doses is less than 1/10 of the LAMA (Maximum Acceptable Annual Limit), monitoring is not justified and should be excluded [27, 29].

There are three types of monitoring for the IOE workplace, which do not necessarily work at the same time: for external radiation, for surface contamination and for air contamination. Likewise, the three types of confirmatory monitoring for IOE are used according to their implementation needs: individual monitoring for external radiation, for internal contamination and for skin and clothing contamination.

CHAPTER 4

MATERIALS AND METHODS

4.1 Study site: Radiopharmacy Centre

The microbiological laboratory where the Albira microPET/SPECT/CT system is installed, which is used for research on small animals, is classified by the facility's radioprotection as a supervised area, which corresponds to an area where, although there is no need to adopt specific protection and safety measures, regular reassessments of occupational exposure conditions must be carried out in order to determine whether the classification of work areas remains adequate and complies with the restrictions and dose limits established by the regulatory authority, CNEN [8, 28].

The following steps were taken to develop this project:

4.2 Stage 1 - Pre-operational workplace monitoring

In this first stage, a portable Radcal ionisation chamber detector, Model 9010 (10x5-1800) (Figure 4.1a), was used to carry out the monitoring, which included analysing the microPET/CT equipment, the operating conditions and the environment of the laboratories involved. It is important to carry out pre-operational monitoring in order to be aware of the background radiation, as well as to check whether there is any influence from other activities carried out in the Radiopharmacy Centre during normal operation. To make the measurements more accurate, a tripod was used (Figure 4.1b) and measurements were taken around the equipment at pre-established distances.

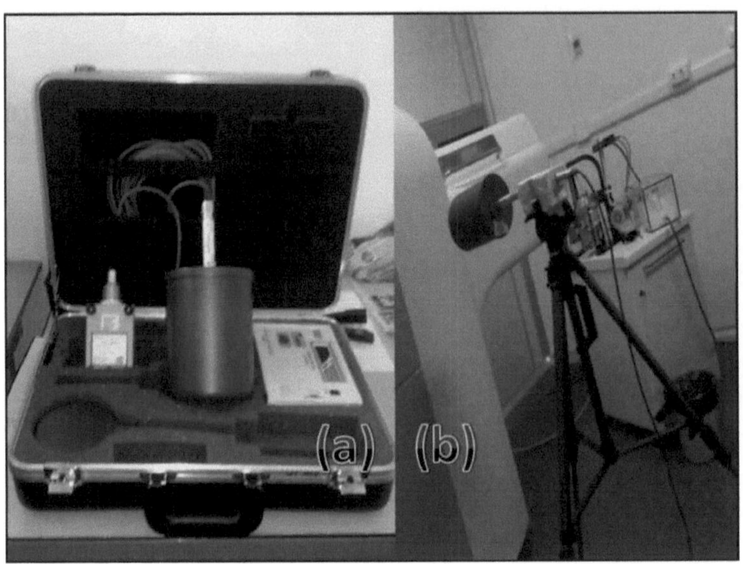

Figure 4.1 (a) Radcal ionisation chamber, Model 9010 (10x5-1800); (b) tripod assembly with the ionisation chamber.

4.3 Stage 2 - Thermoluminescent dosimetry, TLD - Workplace

The level of external radiation at the CR facility was measured using thermoluminescent dosimetry, by placing nine thermoluminescent dosimeters (TL). The dosimeter consists of a plastic detector holder with lead filters and a set of detectors (thermoluminescent tablets of calcium sulphate doped with dysprosium - CaSO4:Dy - with polytetrafluoroethylene).

The TL dosimeters were fixed to the wall at a height of 1.5 metres (referring to the height of the dosimeters' position on the IOEs' chests) from the ground, in locations previously selected according to the sketch shown in Figure 4.2. The monitoring points chosen were: one in the radiopharmaceutical preparation/administration room, two in the microbiological room, three in the room where the microPET/CT is installed, one in the room where the animals are kept, one in the corridor (free area) and one in the biological waste room.

Monitoring the laboratories involved in microPET/CT equipment activities and the surrounding areas will make it possible to detect any deviation from normality (free area) and help control occupational exposures.

Figure 4.2 Arrangement of the nine TL dosimeters in the microPET/CT laboratory.

Data collection: monitoring took place from April 2014 to November 2015, with a monthly exchange rate for another similar dosimeter. The dosimeters were exposed at the nine selected points for 30 days and then sent for reading to IPEN's Thermoluminescent Dosimetry Laboratory (LDT).

During the exchange, the dosimeters were accompanied by a "control dosimeter", which remained in a shield in the LDT for the duration of the dosimeter exposure, in order to accurately determine the dose.The reports of the measurement results, expressed in the physical quantity ambient dose equivalent (mSv) of the nine monitoring points, were provided by the LDT. The air kerma conversion coefficient for the ambient dose equivalent "H*(10)" of 1.14 Sv/Gy was used in the calculations, in accordance with CNEN Regulatory Position 3.01/002:2011 [25].

The ambient dose equivalent values obtained (in mSv) were corrected by the control dosimeters. The method's minimum detection limit is 0.02 mSv [30].

4.4 Stage 3 - Individual monitoring

Estimates of effective dose and dose equivalent are made in accordance with the monitoring programme established in the facility's radiological protection plan.

Occupationally exposed individuals (OEIs) are monitored for external radiation using a thermoluminescent dosimeter (TL) positioned in the chest area.

Control of the incorporation of radioactive material by IOEs is carried out by means of in vivo measurements (whole body counter) every six months. The results of the individual annual effective doses of three workers directly involved in handling microPET/SCPECT/CT between 2013 and 2015 were analysed by consulting their individual dose histories and comparing them with the dose limits established in the CNEN-NN- 3.01 standard [8].

4.5 Stage 4 - Carrying out microPET/CT performance tests

Using the manufacturer's standard simulator, made of acrylic or polymethylmethacrylate (PMMA), the tests were divided into two parts:

Tests for PET: the NEMA NU 4-2008 standard [9] proposes a standardised methodology for evaluating the performance of PETs designed for animal imaging.

The tests carried out in this study are presented below:

I) Spatial resolution: comprises the equipment's ability to distinguish between two points after image reconstruction. For this measurement, the radionuclide used was ^{22}Na (370 kBq on 1/10/12), which is a point source encapsulated in acrylic material (Figure 4.3 (b)). As there is a loss in spatial resolution when the source is moved from the centre of the field of view (FOV) to the edges, the tests were carried out in the centre and at other points in the radial direction, every 5 mm, as shown in Figure 4.3 (a). To make the measurements more precise, a Styrofoam mould the size of the microPET detector's FOV was prepared, making it easier to move the source (Figure 4.3 (c) and (d)).

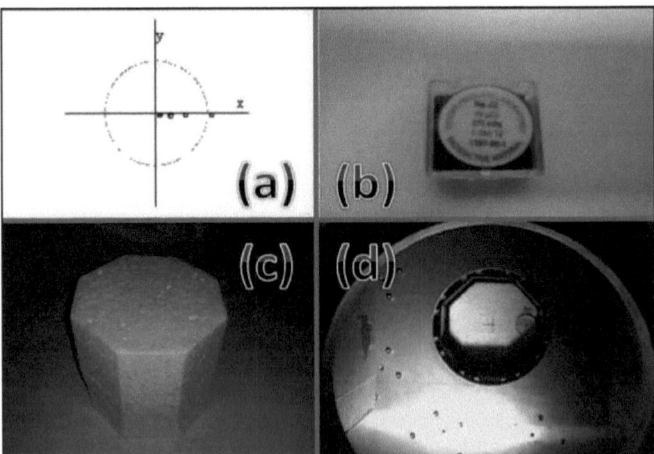

Figure 4.3 (a) Movement of the source in the radial direction, on the x and y axes; (b) $(^{22)}$Na point source encapsulated in acrylic material; (c) Styrofoam mould the size of the FOV of the microPET detector; (d) image of the internal camera, from the centre of the detector's FOV.
centre of the detector's FOV.

II) Scattered fraction, counting losses, and measurements of random events: Each piece of micoPET equipment has different sensitivity to scattered radiation. Figure 4.4 shows the cylindrical simulator used with a hole where the ^{18}F-FDG source with an activity of approximately 51.8 MBq was inserted. Periodic acquisitions were made during the decay of the source (decay curve). This interval lasted 13 hours and 26 images were taken.

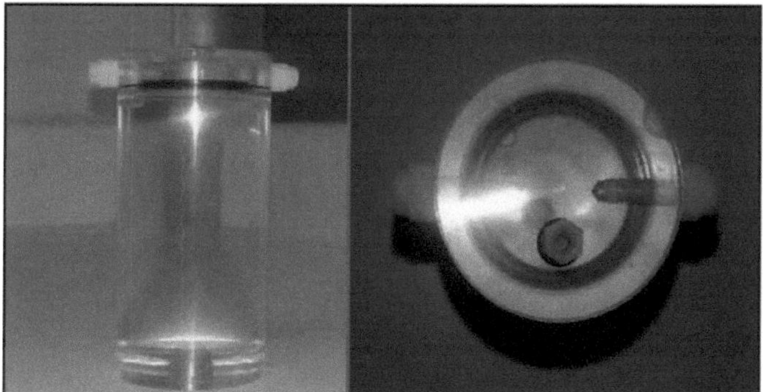

Figure 4.4 Cylindrical PMMA (polymethylmethacrylate) simulator for radionuclide tests.

III) Sensitivity: this is expressed as the rate of true event counts per second detected for a given known source. With the ^{22}Na source in the centre of the scanner, a time period of five minutes was set, which was necessary to collect 10,000 true events (counts).

Test for CT: The leakage radiation test was the only test carried out on the CT module. Using a Radcal Model 9010 ionisation chamber (10x5-1800), measurements were taken on the shielding of the microPET/CT system to check for the possibility of leakage radiation.

4.6 Step 5 - Interpreting the results

Interpretation of the results for the period studied was based on the results presented in the area dosimetry reports supplied by the laboratory (TL) and a comparison of the results of the tests carried out with the standard test protocol pre-established by the equipment manufacturer.

4.7 Estimating Uncertainties

Before presenting the results, it should be noted that the uncertainties of the measurements taken are in accordance with the considerations provided by the Guide to the Expression of Measurement Uncertainty - GUM 2008, translated and adapted by INMETRO, the uncertainties were estimated using the combined standard uncertainty [31, 32].

The combined standard uncertainty is an estimated standard deviation and characterises the dispersion of values that could reasonably be attributed to the measurand. This uncertainty is obtained using equation [31]:

$$u_c = \sqrt{\sum_{i=1}^{n} (u_i)^2}$$

Where uc is the sum of all the uncertainties associated with the measurements taken and ui is the uncertainty of each measurement.

To compose the associated uncertainties, type A and B uncertainties were considered, which are part of probability distributions that indicate the probability of a given measured event being in this uncertainty interval [32].

Type A uncertainties are those estimated by statistical methods. In this case, the standard deviation of the mean was used. Type B uncertainties are related to the limitations and characteristics of the instruments used, such as resolution, calibration factor and others [31, 32]. In this case, the sources of uncertainty used are shown in Table 4.1.

Table 4.1 Sources of type B uncertainties

Sources of uncertainty	Data
Ionisation chamber (calibration)	3,5%
Trena (Resolution)	0.5 mm
Electrometer (Resolution)	0.01 mR/h
Stopwatch (Resolution)	0,01 s

CHAPTER 5

RESULTS AND DISCUSSIONS

5.1 Workplace assessment

The results were divided into three stages: measurements of the room's background radiation level, BG (pre-operational) to check the influence of radioisotope production in normal operation obtained in dose rate mode; measurements of dose rates from the use of microPET/CT operating the PET and PET/CT modules; and evaluation of the ambient equivalent dose reports resulting from the nine selected points.

5.1.2 Measurements of the laboratory's background radiation level (BG)

The average resulting from the 10 measurements of BG dose rates was 0.33 ± 0.10 μSv/h (measurements taken on different days and at different times), with the highest BG value measured during the study period being 0.49 ± 0.05 μSv/h.

5.1.3 MicroPET/CT operating modules: PET and PET/CT

The radionuclide used for the measurements was ^{18}F-labelled fluordesoxyglucose, ^{18}F-FDG. The activities of the ^{18}F-FDG used were: 37 MBq and 59.2 MBq.

a) PET module

Measurements were taken by varying the distance: 30 cm (distance from handling the small animal), 100 cm (distance from the operator moving around the equipment room) and 200 cm (distance from the operator's desk), as shown in Figure 5.1. Five measurements were taken for each distance.

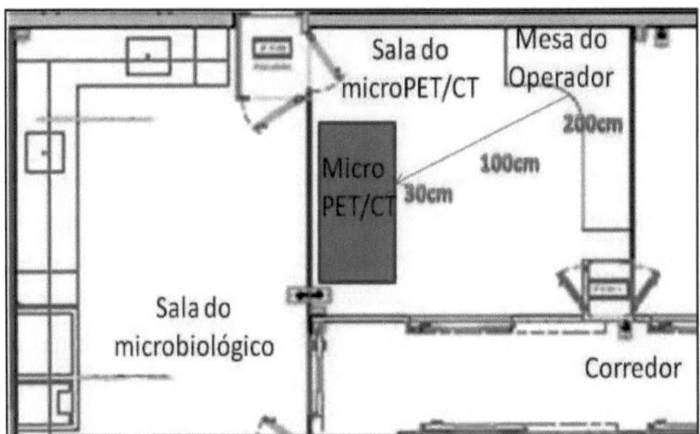

Figure 5.1 Distance of the measurements taken from the door of the leaded glass visor to the operator's table, using the equipment's PET module.

Figure 5.2 illustrates the closed (a) and open (b) leaded glass door of the equipment, where the animal is positioned for the examination or the placement of the simulator with the ^{18}F-FDG used. The dose rate values for distances of 100 and 30 cm [25] are shown in Table 5.1.

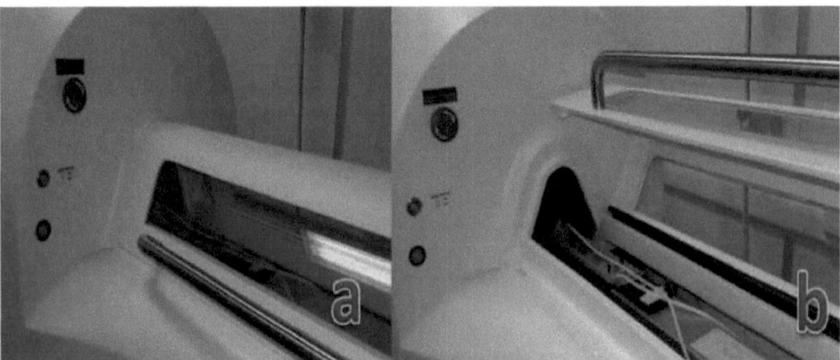

Figura 1.2 Visualisation of the equipment with the leaded glass display door closed (a) and open (b).

Table 5.1 Average dose rate obtained in the PET module

	Distance			
	100 cm		30 cm	
Activity	Lead glass visor door closed (μSv/h)	Lead glass visor door open (μSv/h)	Lead glass visor door closed (μSv/h)	Lead glass visor door open (μSv/h)
37 MBq	1,6 ± 0,3	2,8 ± 0,3	9,0 ± 0,5	16,0 ± 0,6
59.2 MBq	2,5 ± 0,5	4,5 ± 0,5	18,5 ± 3,6	29,0 ± 6,1

The distance of 200 cm was measured only with the leaded glass viewfinder door closed (referring to the operator at the control desk) and an activity of 59.2 MBq, resulting in an average dose rate of 0.11 ± 0.04 μSv/h.

In the PET module, the dose rate with the lead glass viewfinder door open represents the dose rate at which the IOE is exposed when positioning the small animal for data acquisition. The highest value obtained was 29.0 ± 6.1 μSv/h (Tab.5.1).

Assuming that the IOE working on preparing the animal for the examination took 15 minutes a day, on 20 days a month, over 11 months, it could have a maximum dose of 1.6 mSv/year, which is lower than the research limits of over 2.4 mSv/year [8]. Therefore, the shorter the time spent handling the small animal, the lower the effective dose received by the IOE. Taking into account the highest dose rate value of 29.0 ± 6.1 μSv/h (Tab. 5.3), the IOE could not exceed 23 minutes per day when handling the small animal.

b) PET/CT module

The following parameters for the PET/CT module were considered:

C Current of 400μA,

T Voltage 45 kV

R Image resolution around 1000 projections in 25 minutes [15]

The activity of 37 MBq of F^{1}8-FDG was used at a distance of 50 cm.

Figure 5.3 illustrates the monitoring points for the movement of people in the microbiological laboratory room, the corridor and the microPET/CT room.

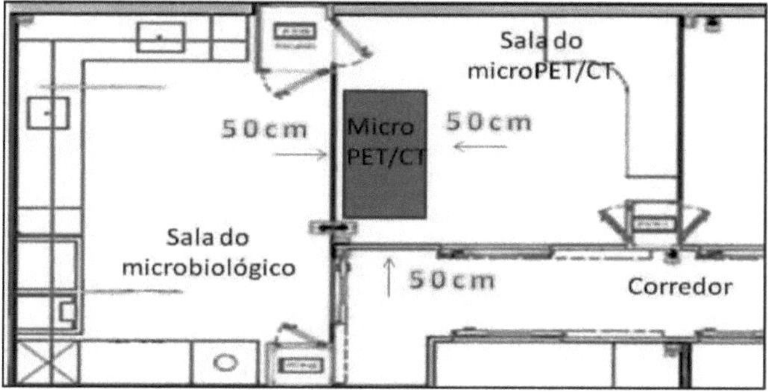

Figure 5.3 Distance of the measurements taken using the PET/CT module of the equipment.

The values measured due to exposure at the selected points (Fig.5.3) were converted into a dose rate [8, 19] and the average dose rate is shown in Table 5.2.

Table 5.2 Average dose rate obtained in the PET/CT module

Monitoring points	Average dose rate μSv/h
MicroPET/CT room	0,29 ± 0,02
Corridor	0,17 ± 0,01
Microbiological room	0,14 ± 0,02

In the PET/CT module, the dose rates were measured from the lining of the equipment and the results were lower than the values measured in the PET module, which had the lead glass viewfinder door as a reference.

5.1.3 Evaluation of the ambient dose equivalent

The area dosimetry reports for the nine monitored points were evaluated from April 2014 to November 2015. The values obtained for the ambient dose equivalent (in mSv) were subtracted from a control dosimeter, shown in Table 5.3 [30].

Table 5.3 Results of the ambient dose equivalent (in mSv) over the period studied.

		Ambient dose equivalent H*(10) (mSv)									
		Dosimeter									
	Month S%	*Control*	**1**	**2**	**3**	**4**	**5**	**6**	**7**	**8**	**9**
2014	**April**	*0,16*	0,33	0,21	0,21	0,2	**0,25**	0,29	0,19	0,18	0,17
	s%	*2,5*	3,8	2,4	1,2	1,4	8,4	1,9	8,3	39	3,3
	May	*0,18*	0,16	0,1	0,09	0,09	**0,15**	0,14	0,08	0,11	0,09
	s%	*4,1*	1,1	5,6	6,1	2	5,7	1,5	6,5	1,9	2,6
	June	*0,18*	0,28	0,1	0,1	0,09	**0,23**	0,22	0,09	0,1	0,08
	s%	*8,1*	2,8	3,4	1,3	13	6,3	12	13	9,5	6,9
	July	*0,19*	0,26	0,14	0,15	0,12	**1,52**	0,2	0,1	0,16	0,12
	s%	*7,9*	11,3	2,8	2,1	3,7	12,8	7,8	4,4	10,5	3,3
	August	*0,31*	0,38	0,15	0,13	0,11	**0,49**	0,15	0,1	0,13	0,1
	s%	*2,6*	*	6,1	6,5	2,6	11	8,4	2,5	9,1	5,3
	September	*0,13*	0,27	0,13	0,09	0,13	**0,2**	0,16	0,08	0,1	0,11
	s%	*4,5*	10	5,2	9,5	1,9	3,3	9	6,2	3	3,1
	October	*0,21*	0,17	0,09	0,1	0,04	**0,38**	0,11	0,05	0,16	0,15
	s%	*4,1*	5,6	4,8	8,9	3,6	9,1	4,2	0,8	1,3	5,5
	November	*0,29*	0,13	0,1	0,09	0,08	**0,8**	0,11	0,04	0,11	0,06
	s%	*8*	8,1	5,6	1,5	5,2	22	4,8	2,1	5,2	7,6
	December	*0,2*	0,15	0,08	0,09	0,03	**0,33**	0,09	0,04	0,14	0,13
	s%	*4,1*	5,6	4,8	8,9	3,6	9,1	4,2	0,8	1,3	5,5
	January	*0,12*	0,2	0,16	0,13	0,15	**0,28**	0,18	0,13	0,17	0,15
	s%	*5,2*	6	7,5	2,3	2,9	8,7	3,2	1,5	2,6	2,9
	February	*0,14*	0,19	0,09	0,08	0,12	**0,26**	0,2	0,08	0,12	0,08
	s%	*3,3*	4,1	4,5	3,7	3,2	3,1	15	5,9	7	3,1
	March	*0,4*	0,12	0,05	0,02	0,03	**0,2**	0,15	0,03	0,06	0,04
	s%	*5,3*	7,2	3,4	2,3	3,7	3,7	8,9	4,6	5,7	5
	April	*0,4*	0,08	0,03	0,04	0,04	**0,26**	0,14	0,04	0,03	0,03

	s%	*5,2*	4,5	1,2	2,6	2,4	21	11	5,7	11	4,6
	May	*0,28*	0,08	0,07	0,04	0,1	**0,18**	0,17	0,05	0,04	0,05
	s%	*1,5*	7,6	0,4	4,7	18	6,9	16	5,2	0,6	2,3
	June	*0,19*	0,06	0,07	0,04	0,05	**0,29**	0,13	0,05	0,05	0,05
	s%	**7**	3,4	5,1	1,9	4,6	16	6,3	4,3	7,2	2,7
	July	*0,22*	0,06	0,06	0,05	0,05	**0,19**	0,12	0,05	0,07	0,05
	s%	*6,1*	2,6	4	13	1,5	2,4	3,8	3,3	12	2,8
	August	*0,25*	0,12	0,08	0,05	0,04	**0,12**	0,19	0,08	0,05	0,05
	s%	*8,9*	3,4	8,4	4,6	3,8	11	7,7	2,6	7,7	2,4
	September	*0,16*	0,2	0,11	0,12	0,1	**0,93**	0,18	0,1	0,13	0,16
	s%	*3,2*	8,8	3,1	6,3	7,9	8,5	14	16	3,9	18
2015	**October**	*0,21*	0,2	0,11	0,09	0,08	**0,58**	0,16	0,19	0,08	0,09
	s%	*8,1*	13	7,3	6,4	6,3	18	11	24	4	8
	November	*0,14*	0,32	0,24	0,19	0,2	**0,49**	0,32	0,33	0,22	0,21
	s%	*3,3*	12	3,3	8	0,7	4,5	17	20	3,5	6,6

* Standard deviation greater than 20%

S% percentage standard deviation of the three detectors present in the same dosimeter [23].

The results in Table 5.3 show that the dosimeter located at point 5 (radiopharmaceutical preparation/administration room) was the one with the highest dose. After investigation, the possible reasons were the handling of a larger quantity of radiopharmaceuticals and also the storage of unsealed radioactive sources on site.The result of the average ambient dose equivalent (mSv) resulting from the nine points monitored with a TL dosimeter are shown in Table 5.4.

Table 5.4 Average ambient dose equivalent (mSv) over the study period

Dosimeter	Average Ambient Dose Equivalent (mSv)
1	0,19 ± 0,09
2	0,11 ± 0,05
3	0,10 ± 0,05
4	0,09 ± 0,05
5	0,41 ± 0,34
6	0,17 ± 0,06
7	0,10 ± 0,07
8	0,11 ± 0,05
9	0,10 ± 0,05

It can be seen in Table 5.4 that the average ambient dose equivalent at point 5 is higher than at the other points; the value is within the limits of a supervised area which is in the range of 0.08 mSv/month to 0.5 mSv/month [8].

The microPET/CT facilities are classified as a supervised area, so attention must be paid to the doses so as not to exceed the annual limits. In July 2014, point 5 had a dose of 1.52 mSv which led to an investigation; it is not possible to specify the quantity of radiopharmaceuticals that entered the laboratory or the increase in the number of tests carried out [8, 24].

As of November 2014, two forms have been implemented for recording users, in order to have greater control over the activity handled by radiopharmaceuticals and the storage of unsealed radioactive sources in the room. One of the forms, called "Register of the use of radiopharmaceuticals in the MicroPET/CT laboratory", [Annex A] records the entry and exit of material in the radiopharmaceutical preparation/administration room and in the MicroPET/CT room. Another form records the use of the equipment, called "Albira MicroPET/CT equipment use record" [Appendix B].

5.2 Individual Monitoring

All the IOEs at CR IPEN-CNEN/SP directly involved in studies or research related to processes involving the microPET/CT facilities follow the recommendations of Radioprotection. The result of the effective doses of the three IOEs presented in Table 5.6 shows that the dose values for the years 2013 to 2015 are lower than or practically equal to the record level for annual individual monitoring, which is equal to 2.4 mSv, as established by the CNEN-NN-301 standard [8].

Table 5.5 Effective dose of IOEs in the period studied

IOE	Effective Dose (mSv) 2013	Effective Dose (mSv) 2014	Effective Dose (mSv) 2015
A	2,24	2,46	2,40
B	2,40	2,20	2,40
C	*	2,20	2,40

*IOE has no monitoring data, it wasn't in the institution.

Activities with microPET/CT began in November 2013, but the on-site study started in April 2014. Table 5.5 shows that there was no significant increase in individual doses.

5.3 Performance tests on microPET/CT

5.3.1 PET module

By testing the Albira microPET/CT equipment in accordance with the NEMA NU 4-2008 standard [9], it was possible to verify the operation of the PET acquisition mode. The tests were adapted to the conditions of the experimental setups.

The purpose of carrying out performance tests on the microPET is to check that the values obtained are within those proposed by the manufacturer, i.e. that during this period the equipment still maintains its conditions of use.

The images generated were visualised using the AMIDE programme, as it is an available and useful tool for visualisation, as well as analysing and registering volumetric medical image data sets

[33, 34].

5.3.1.1 Spatial resolution

Figure 5.4 schematically shows the source arrangement and position for the resolution measurements. The ^{22}Na source was moved from the centre of the field of view (FOV) to the edges on the x and y axes, with z constant at the centre of the FOV between the two detection rings. The aim was to characterise the widths of the images reconstructed from a punctiform scattering function, measured as the full width at half-maximum amplitude *(FWHM) and* the full width at one *tenth-maximum amplitude (FWTM).* The lower the value, the better the resolution of the equipment [9, 17]. The Amide programme was used to calculate the FWHM and FTHM in three sections: transverse, coronal and sagittal (Fig. 5.5).

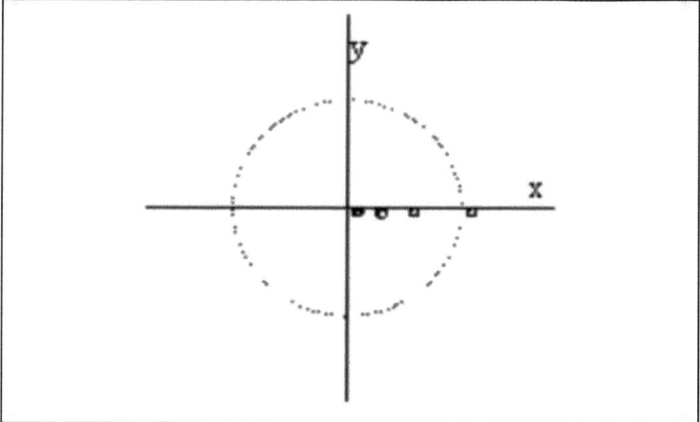

Figure 5.4 Representation of the movement of the ^{22}Na source from the centre to the edges.

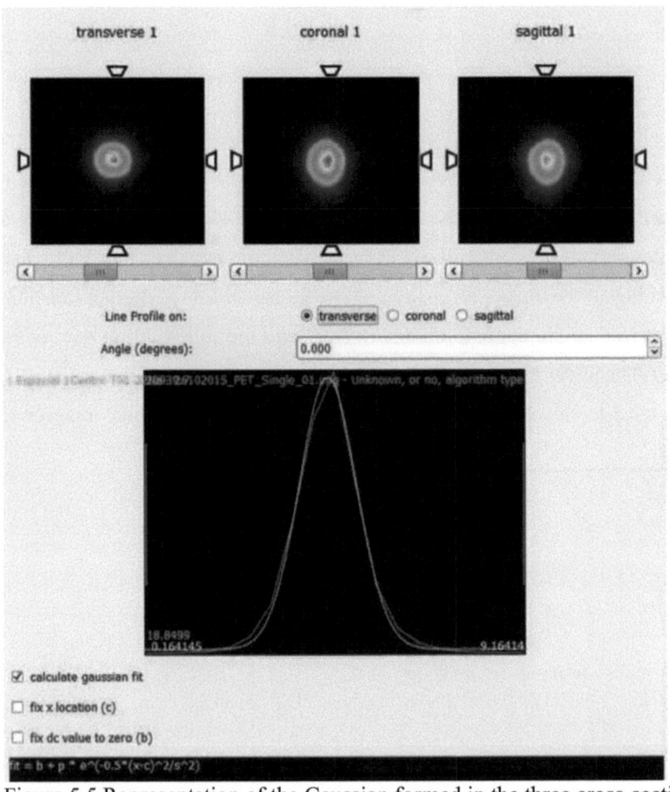

Figure 5.5 Representation of the Gaussian formed in the three cross-sections, coronal and sagittal.

Tables 5.6 to 5.10 show the values obtained with their respective FWHM and FWTM uncertainties.

Table 5.6 FWHM (mm) and FWTM (mm) values at position 0 in "X" and "Y" (mm)

TRANSVERSAL			**CORONAL**		**SAGITTARIUS**	
Position 0 in	**FWHM**	FWTM	**FWHM**	FWTM	**FWHM**	FWTM
"X" and "Y"	**1,6157±0,000080284**	2,9447±0,00014633	**1,6157±0,000080284**	2,9447±0,00014633	**1,5881±0,000080024**	2,8945±0,000014585

Table 5.7 FWHM (mm) and FWTM (mm) values in the right "X" positions (mm)

TRANSVERSAL			**CORONAL**		**SAGITTARIUS**	
X"	FWHM	FWTM	FWHM	FWTM	FWHM	FWTM

positions right						
5	**1,7326±0,000095286**	3,1578±0,0001736	**1,7326±0,0000955286**	3,1578±0,00017367	**1,648±0,00010161**	3,0037±0,0001852
10	**1,8812±0,00011437**	3,4286±0,00020844	**1,8812±0,00011437**	3,4286±0,00020844	**1,5816±0,000093013**	2,8827±0,00016953
15	**2,2208±0,00017164**	4,0476±0,00031284	**2,2208±0,00017164**	4,0476±0,00031284	**1,7061±0,00012488**	3,1096±0,00022762
20	**2,6405±0,00022777**	4,8126±0,00041514	**2,6405±0,00022777**	4,8126±0,00041514	**1,8791±0,00016891**	3,4249±0,00030786
25	**2,46447±0,00026897**	4,8203±0,00049023	**2,46447±0,00026897**	4,8203±0,00049023	**1,956±0,00019988**	3,565±0,0003643
30	**2,9585±0,00037276**	5,3923±0,00067939	**2,9585±0,00037276**	5,3923±0,00067939	**1,9252±0,00024509**	3,5089±0,00044671
35	**3,1578±0,00041942**	5,7554±0,00076445	**3,1578±0,00041942**	5,7554±0,00076445	**1,9021±0,00023352**	3,4667±0,00042561
40	**5,3251±0,085277**	9,7056±0,15543	**5,3251±0,085277**	9,7056±0,15543	**2,7766±0,00012515**	5,0606±0,00022811

Table 5.8 FHM (mm) and FWTM (mm) values in the left "X" positions (mm)

TRANSVERSAL			CORONAL		SAGITTARIUS	
Left "X" positions	**FWHM**	FWTM	**FWHM**	FWTM	**FWHM**	FWTM
40	**4,5493±0,096803**	8,2916±0,17643	**4,5493±0,096803**	8,2916±0,17643	**2,7732±0,00014987**	5,0544±0,00027315
35	**2,8017±0,00032275**	5,1065±0,00058825	**2,8017±0,00032275**	5,1065±0,00058825	**1,8122±0,0002116**	3,303±0,00038566
30	**2,8771±0,00028896**	5,2439±0,00052666	**2,8771±0,00028896**	5,2439±0,00052666	**2,0134±0,00021579**	3,6697±0,0003933
25	**2,6551±0,00027052**	4,8393±0,00049305	**2,6551±0,00027052**	4,8393±0,00049305	**1,8817±0,00018846**	3,4295±0,00034349
20	**2,4659±0,00020304**	4,4943±0,00037007	**2,4659±0,00020304**	4,4943±0,00037007	**1,8132±0,00015094**	3,3048±0,0002751
15	**2,1411±0,00014533**	3,9024±0,00026489	**2,1411±0,00014533**	3,9024±0,00026489	**1,7685±0,00013151**	3,2232±0,00023968
10	2,0109±0,00012741	**3,665±0,00023222**	2,0109±0,00012741	**3,665±0,00023222**	1,7945±0,00011565	3,2707±0,00021079
5	**1,665±0,000096692**	3,0347±0,00017623	**1,665±0,000096692**	3,0347±0,00017623	**1,6291±0,000092812**	2,9692±0,00016916

Table 5.9 FWHM (mm) and FWTM (mm) values in the upper "Y" positions (mm)

TRANSVERSAL			CORONAL		SAGITTARIUS	
Upper "Y" positions	**FWHM**	FWTM	**FWHM**	FWTM	**FWHM**	FWTM
5	**1,7117±0,00010164**	3,1198±0,00018525	**1,7117±0,00010164**	3,1198±0,00018525	**1,7113±0,00010034**	3,119±0,00018289
10	**1,6827±0,0001075**	3,0669±0,00019592	**1,6827±0,0001075**	3,0669±0,00019592	**1,9945±0,00012208**	3,6352±0,0002225
15	**1,8522±0,00014508**	3,3758±0,00026443	**1,8522±0,00014508**	3,3758±0,00026443	**2,3366±0,00018807**	4,2588±0,00034277
20	**1,8116±0,00015317**	3,3019±0,00027916	**1,8116±0,00015317**	3,3019±0,00027916	**2,4696±0,00019804**	4,5011±0,00036096
25	**1,8754±0,00017824**	3,4182±0,00032487	**1,8754±0,00017824**	3,4182±0,00032487	**2,6414±0,00024881**	4,8146±0,00045349
30	**1,9472±0,00020493**	3,5491±0,0003735	**1,9472±0,00020493**	3,5491±0,0003735	**2,7093±0,0002743**	4,938±0,00050008
35	**1,9433±0,0002567**	3,5419±0,00046786	**1,9433±0,0002567**	3,5419±0,00046786	**3,0159±0,0004041**	5,4969±0,00073652

40	**2,4831±0,00013 169**	4,5258±0,00024 002	**2,4831±0,00013 169**	4,5258±0,00024 002	**7,8962±0,11053**	14,392±0,2146

Table 5.10 FWHM (mm) and FWTM (mm) values in the lower "Y" positions (mm)

TRANSVERSAL			**CORONAL**		**SAGITTARIUS**	
Lower "Y" positions	**FWHM**	FWTM	**FWHM**	FWTM	**FWHM**	FWTM
40	**2,983±0,000143 07**	5,4369±0,0002 6077	**2,983±0,000143 07**	5,4369±0,0002 6077	**4,5779±0,0905 95**	8,3437±0,16512
35	**1,964±0,000262 24**	3,5795±0,0004 7795	**1,964±0,000262 24**	3,5795±0,0004 7795	**3,0717±0,0004 1087**	5,5985±0,00074 886
30	**1,9126±0,00020 773**	3,486±0,00037 861	**1,9126±0,00020 773**	3,486±0,00037 861	**2,7273±0,0002 8832**	4,9708±0,00052 55
25	**1,8667±0,00019 096**	3,4023±0,0003 4805	**1,8667±0,00019 096**	3,4023±0,0003 4805	**2,651±0,00026 836**	4,8317±0,00048 912
20	**1,797±0,000154 07**	3,2572±0,0002 8081	**1,797±0,000154 07**	3,2572±0,0002 8081	**2,4577±0,0001 8431**	4,4795±0,00033 593
15	**1,6965±0,00012 182**	3,092±0,00022 203	**1,6965±0,00012 182**	3,092±0,00022 203	**2,2171±0,0001 552**	4,0408±0,00028 286
10	**1,6229±0,00011 361**	2,9579±0,0002 0707	**1,6229±0,00011 361**	2,9579±0,0002 0707	**2,0752±0,0001 2503**	3,7823±0,00022 788
5	**1,6195±0,00009 2094**	2,9518±0,0001 6785	**1,6195±0,00009 2094**	2,9518±0,0001 6785	**1,7112±0,0001 0782**	3,1189±0,00019 651

The differences in spatial resolution at each point for the three cross-sections, coronal and sagittal are illustrated in graphs 5.1 to 5.4 on the x and y axes. The uncertainties of the measurements are not shown in the graphs; most have values of less than 0.01% of the measurement.

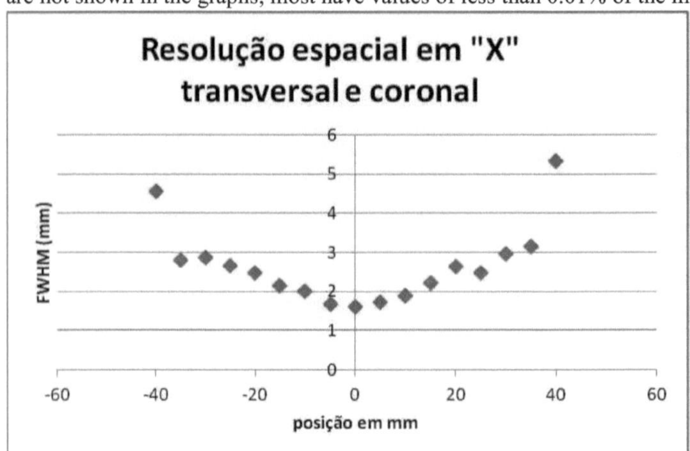

Graph 5.1 Transverse and coronal X-ray spatial resolution

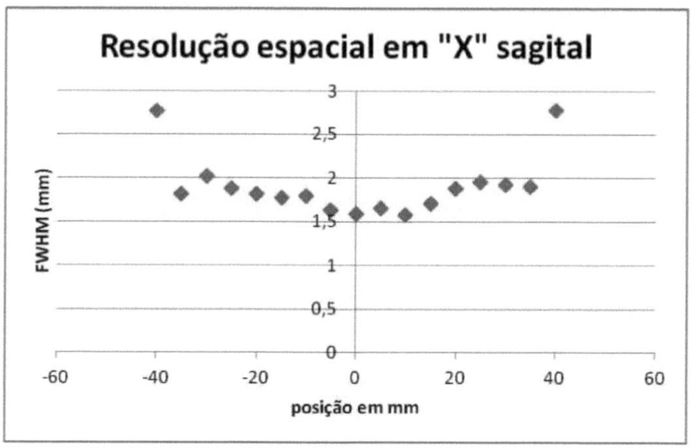

Graph 5.2 Sagittal "X" spatial resolution

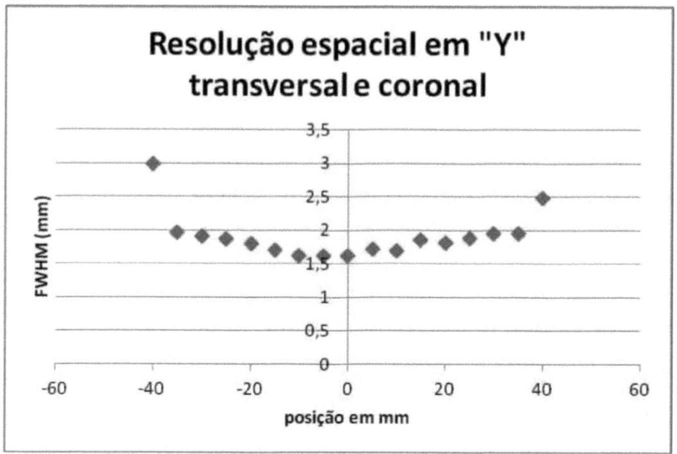

Graph 5.3 Transverse and coronal "Y" spatial resolution

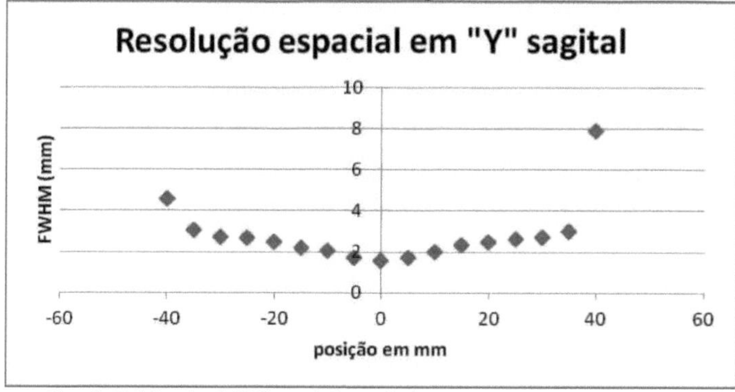

Graph 5.4 Sagittal "Y" spatial resolution

In the transverse and coronal plane in the "x" axis, the variation in spatial resolution was 1.61 to 5.32 mm. In the sagittal "x" plane, the spatial resolution ranged from 1.58 to 2.77 mm.

In the transverse and coronal plane in the "y" axis, the variation in spatial resolution was 1.61 to 2.98 mm. In the sagittal "x" plane, the variation in spatial resolution ranged from 1.58 to 7.89 mm.

The graphs show that as the source moved from the centre to the edges there was a loss of resolution. In the centre of the detector, position zero, it was around 1.6 mm, which is 20% more than the manual value of 1.2 mm [15]. However, in other studies the spatial resolution of the Albira MicroPET/SPECT/CT in the centre of the detector was 1.5 mm [35, 36].

5.3.1.2 Sensitivity

Sensitivity was also measured with the ^{22}Na point source. The data was calculated only in the centre of the field of view (FOV). The source was positioned on the bed and introduced up to the centre of the detector as shown in the images in Figure 5.6.

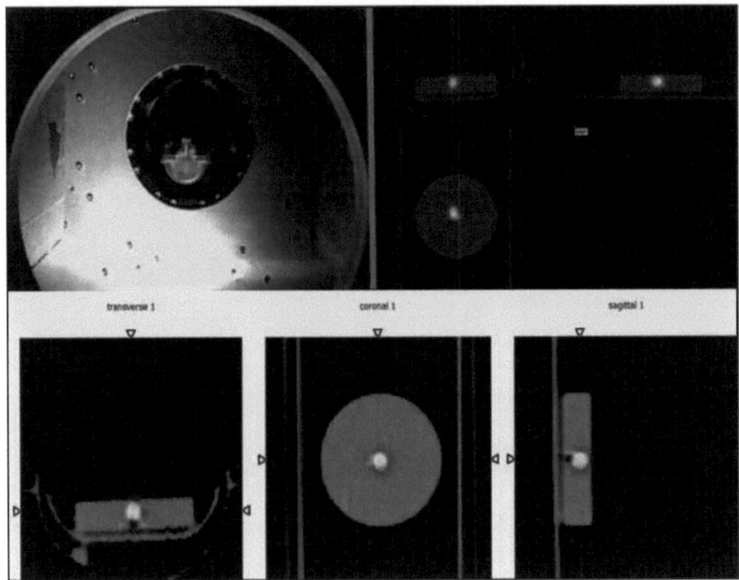

Figure 5.6 Image generated from the sensitivity test.

Since the "sinogram" data (which consists of the recording of coincidence events on a PET scanner in which the response lines are described as a function of angular orientation and position relative to the centre of the field of view) [37] was not accessible (due to a lack of understanding of the programming codes), the procedure as set out in NEMA [9] was not carried out.

The absolute sensitivity (%), "SA" was then calculated as [37]:

SA = (St * 100) / (A * 0.906)

Where:

St - is the variation of event count with source - event count without source

A - is the activity of the certified source (Bq) supplied by the manufacturer and 0.906 is the branching fraction of ^{22}Na (a standard for this source, for this use).

The half-life of ^{22}Na is 2.6 years and the "St" is 6.66 k-events.

The value obtained in the absolute sensitivity test carried out on the centre of the source was 4.83%, which is close to the value given by the manufacturer, which is 5% [15].

5.3.1.3 Spread fraction, counting losses and measures of random events

Two days of measurements were carried out, with each measurement lasting 13 hours. The decay time of the source was also observed. On each day, 26 images were taken with an acquisition time of 300 seconds and an interval between acquisitions of 1800 seconds. The results are shown in Table 5.11 and illustrated in Graphs 5.5 to 5.7.

Table 5.11 Results of the measurements taken during the decay of the ^{18}F-FDG source with an activity of 51.8 MBq

Measure 1		Measure 2		Calculated	
Time (s)	Counting	Time (s)	Counting	Time (s)	Activity (MBq)
148,71	0	148,88	0	0	0
1960,9	98,06	1961,6	114,39	1800	42,86
3774,4	168,37	3775,3	189,32	3600	35,47
5588	264,22	5589	291,32	5400	29,35
7401,6	385,01	7401,2	412,68	7200	24,28
9215,3	530,99	9214	570,86	9000	20,09
11029	705,59	11027	772,87	10800	16,63
12844	723,68	12841	777,39	12600	13,76
14657	695,88	14654	764,07	14400	11,39
16469	664,83	16467	751,35	16200	9,42
18281	655,19	18280	751,99	18000	7,8
20095	629,35	20093	742,29	19800	6,45
21908	608,72	21906	715,4	21600	5,34
23722	573,47	23718	713,01	23400	4,42
25535	538,06	25531	674,82	25200	3,65
27348	486,54	27344	628,95	27000	3,02
29161	431,45	29158	595,32	28800	2,5
30974	359,02	30970	512,21	30600	2,07

32788	279,16	32783	454,82	32400	1,71
34600	186,38	34596	350,58	34200	1,42
36413	92,35	36409	231,69	36000	1,17
38227	18,47	38221	76,35	37800	0,97
40039	1,25	40034	21,4	39600	0,8
41852	0,18	41847	2,25	41400	0,66
43666	0	43660	0	43200	0,55
45478	0	45472	0	45000	0,46

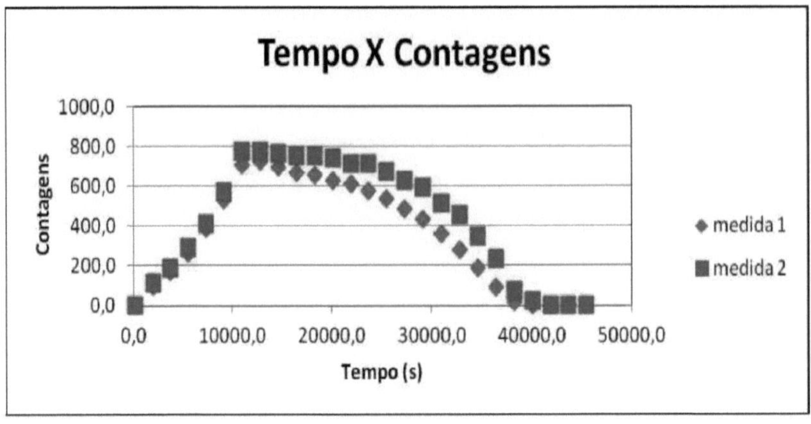

Graph 5.5 Time x Counts

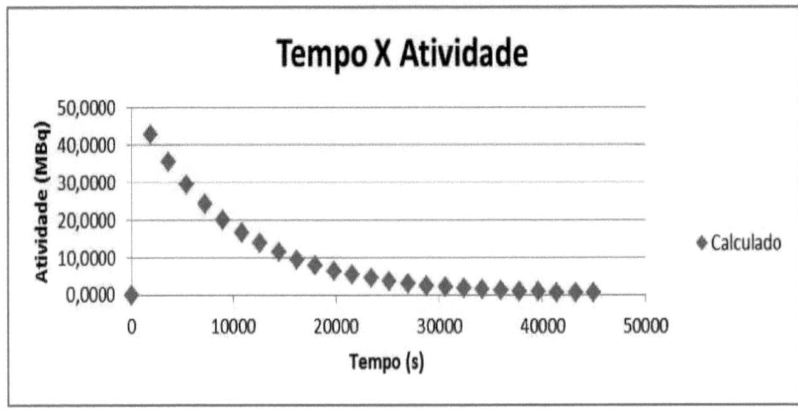

Graph 5.6 Time x Activity

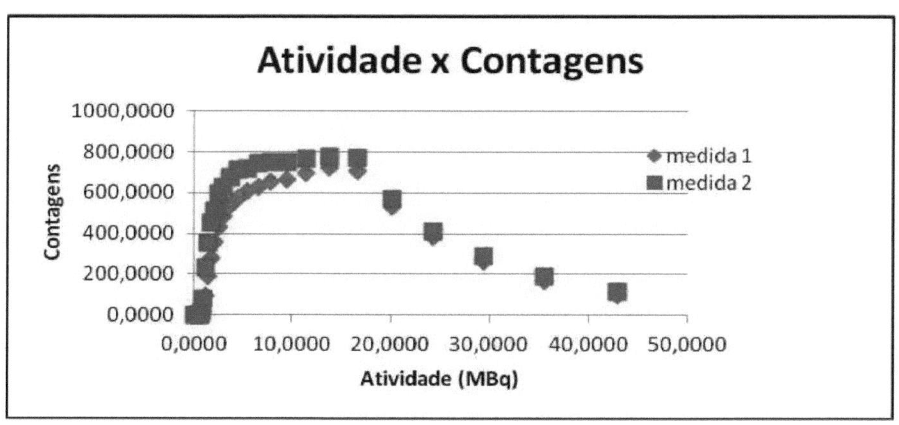

Graph 5.7 Activity x Counts

Comparing graphs 5.5 and 5.6, it can be seen that the ^{18}F-FDG decay curve behaves differently to the one calculated. In graph 5.5, the curve rises, representing the detector's inability to count the radiation, due to the saturation of the crystal as a result of the high activity of ^{18}F-FDG used. Graph 5.7 shows a rise, then the curves reach a plateau (9.4 to 16.6 MBq) and then fall. This plateau represents the interval in which the detector is operating at its maximum limit. Therefore, in order for the detector to count in such a way that the crystal does not saturate, it is important to consider the ideal operating activity of the device, i.e. less than 9.4 MBq for ^{18}F-FDG [36, 37, 38].

5.3.2 CT Module - Leakage Radiation Test

To check for possible radiation escaping from the equipment's shielding, 12 points were selected according to the manufacturer's acceptance criteria, where none of these points can exceed a dose rate of 5 μSv/h [14]. The selected points are shown in Fig. 5.6.

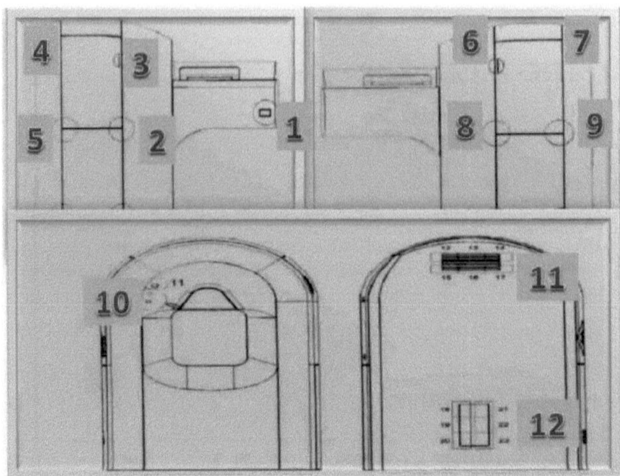

Figure 5.7 Selected points on the shield for leakage radiation measurements.

Where:

Point1 represents the place where the anaesthesia system enters;

Points 2, 3, 4, 5, 6, 7, 8 and 9 represent the equipment's port junctions;

Point10 represents the button for switching the device on and off;

Point11 the air outlet and finally,

Point 12 represents the wires that connect the equipment to the mains.

The results of these measurements are shown in Table 5.12.

Table 5.12 Dose rate at the 12 shield points

Points on the armour	**Dose rate μSv/h**
1	0,83 ± 0,06
2	1,15 ± 0,06
3	0,98 ± 0,12
4	1,08 ± 0,08
5	1,09 ± 0,09
6	0,37 ± 0,03
7	0,41 ± 0,02
8	0,34 ± 0,06
9	0,34 ± 0,06
10	1,11 ± 0,04
11	0,94 ± 0,07
12	1,16 ± 0,06

The results of the dose rate measurements at selected points on the shield do not indicate radiation leakage in excess of the acceptance limit of 5 |jSv/h, as established by the manufacturer [14].

CHAPTER 6

CONCLUSIONS

The microPET/CT laboratory does not have a daily routine, but this work has shown that, in relation to occupational exposure from the use of the equipment, self-shielding is effective and there is no need to change the physical structure of the laboratory, since at no point did the radiation dose exceed the limits indicated in the standards.

The average background radiation obtained was 0.33±0.10 μSv/h and the dose rates during the microPET/CT operation were less than 5 μSv/h for a distance of 1 m.

In July 2014, the value obtained for the H*(10) ambient equivalent dose was 1.52 mSv/month, which led to concern about the time (occupancy factor) that the IOEs would be handling the radiopharmaceuticals.

The results of the assessment of radiation levels in the MicroPET/CT laboratory areas show that the dose estimates in relation to CNEN standards are adequate. All the points were considered satisfactory, showing that the shielding of the equipment is adequate and in accordance with the classification of the supervised and free areas of the facility.

The average annual effective dose of the monitored IOEs did not exceed the registration level of 2.4 mSv/year.

With regard to the tests carried out, it was possible to get an idea of some of the device's parameters, which could help with further studies in radiopharmacy: such as the best energy range to use and the region of best resolution for the device. These tests can also be improved to form part of a set of tests for quality control of the device.

Currently, SPECT has been installed and there will probably be an increase in the number of radiopharmaceuticals that can be used in the laboratory and the number of people involved in the activities, as well as the levels of radiation emitted by microPET/SPECT/CT, highlighting the importance of continuing to monitor the workplace.

The importance of strictly following the principles of radioprotection should be emphasised, as this is a laboratory where unsealed radioactive sources are handled.

BIBLIOGRAPHICAL REFERENCES

[1] SILVA M. I. B.; ***"Physical Characterisation of a PET/CT Imaging System"***; Dissertation (Master's Degree), Technical University of Lisbon; Portugal; 2008

[2] BETTINARDI V., DANNA M., SAVI A., LECCHI M., CASTIGLIONI I., GILARDI M. C., BAMMER. H, LUCIGNANI G., FAZIO F.; *"Performance Evaluation of the **New Whole-Body** PET/CT Scanner: Discovery ST"*; European Journal of Nuclear Medicine and Molecular Imaging; Jun,31(6):867-81; 2004

[3] ROBILOTTA C.C.; *"Positron Emission Tomography: A New Modality in Brazilian Nuclear Medicine"*, Rev. Panam Salud Publica. 2006; 20(2/3):134-42, 2006

[4] KUBO A. L. S. L.; *"Study of External Occupational Exposure in Nuclear Medicine Services in Rio de Janeiro"*; IRD; Rio de Janeiro; 2011

[5] BEYER T., TOWNSEND D. W., BRUN T., KINAHAN P. E., CHARRON M., RODDY R., JERIN J., YOUNG J., BYARS I. and NUTT R.; *"A Combined PET/CT Scanner for Clinical Oncology"*; The Journal of Nuclear medicine; Vol 41, N° 8;
2000

[6] PEREIRA N. J. S.; *"MicroPET Scanner Characterisation"*; Dissertation (Master's Degree), Technical University of Lisbon; Portugal; 2008

[7] MUSTAFAD et al; ***"Radiation Doses to Technologists Working with 18F-*** *FDG in a PET Centre with high Patient Capacity"*; Nukleonika; 55(1)107-112;
Turkey; 2010

[8] NATIONAL COMMISSION ON NUCLEAR ENERGY; *"Basic Guidelines for Radiological Protection"*; Standard CNEN-NN-3.01; Resolution CNEN 164/14; (Amendment of item 5.4.3.1); Rio de Janeiro; Publication DOU: 11/03/2014

[9] NATIONAL ELECTRICAL MANUFACTURERS ASSOCIATION; *"Performance Measurements of Small Animal Positron Emission Tomographs"*; NEMA NU 4-2008; USA; 2008

[10] INTERNATIONAL ATOMIC ENERGY AGENCY; *"Nuclear Medicine Resources Manual"*; Vienna; 2006

[11] TAUHATA L., SALATI I.P.A., PRINZIO R. D.; *"Radioprotection and Dosimetry: Fundamentals"*; Institute of Radioprotection and Dosimetry/CNEN; 5 Rev.

[12] BRAZILIAN SOCIETY OF BIOLOGY NUCLEAR MEDICINE AND MOLECULAR IMAGING; *"The Differences Between PET/CT and PET Scan"*; Source: <http://portaldaradiologia.com>, accessed on 06/05/2013

[13] BALTAZAR C.; *"Avianca Begins Transport of FDG-^{18}F **that Impacts on the Diagnosis and Management of** Cancer **Patients**"*; Source: <www.avianca.com.br>; Accessed on 05/04/2013

[14] SQUAIR P.L; *"Equipment Installation (IJPET)"*; Albira Imaging System; GPD - DIRF R&D Management, IPEN/CNEN; São Paulo; 03/2013

[15] BRUKER CORPORATION; *"Bruker Albira Imaging System User Manual"*; IB5438511 Rev. C 12/12; 2012

[16] SQUAIR P.L; ***"Equipment** Installation, **UPGRADE - CTfaPET-CT)"**;* Albira Imaging System; GPD - DIRF R&D Management, IPEN/CNEN; São Paulo; 11/2013

[17] SÁ L.V; *"Quality Control in Positron Emission Tomography"*; Thesis (PhD), COPE/UERJ; Rio de Janeiro; 2010

[18] INTERNATIONAL ATOMIC ENERGY AGENCY; *"Quality Assurance for **PET and PET/CT** Systems";* HUMAN HEALTH SERIES N° 1; Vienna; 2009

[19] NATIONAL COMMISSION ON NUCLEAR ENERGY; *"Radiological Safety and Protection Requirements for Nuclear Medicine Services"*; CNEN Standard NN 3.05; CNEN Resolution 159/13, December/2013; Rio de Janeiro; 2013

[20] SECRETARIA DE VIGILÂNCIA SANITÁRIA; *"Diretrizes de Proteção Radiológica em Radiodiagnóstico Médico e Odontológico"*; Ministério da Saúde, Anvisa; Portaria/MS/SVS n° 453; de 01 de junho de 1998; DOU de 02/06/98

[21] NATIONAL ELECTRICAL MANUFACTURERS ASSOCIATION; *"About the National Electrical Manufacturers Association"*; Source: <http://www.nema.org>; Accessed on 02/03/16

[22] PAÚRA C.L. and DANTAS B.M.; ***"Evaluation of Occupational Exposure in the Use of** ^{18}F in Positron Emission Tomography Examinations"*; XI Brazilian Congress of Medical Physics - Proceedings; Ribeirão Preto; 2006

[23] SANTANA P. C.; OLIVEIRA P. M. C.; MAMEDE M.; SILVEIRA M. C.; AGUIAR P.; REAL R.V.; SILVA T. A.; *"Ambient Radiation Levels in Positron Emission Tomography/Computed Tomography (PET/CT) Imaging Centre"*; Radiologia Brasileira; vol. 48/n° 1; 2015

[24] UNITED NATIONS SCIENTIFIC COMMITTEE ON THE EFFECTS OF ATOMIC RADIATION; *"Sources and Effects of Ionizing Radiation"*; UNSCEAR 2008; Report Vol.I; 2010

[25] NATIONAL COMMISSION ON NUCLEAR ENERGY, ***"Weighting** Factors **for** Radiological **Protection Quantities"***; Regulatory Position 3.01/002:2011; Rio de Janeiro; Brazil; 2011

[26] INTERNATIONAL COMMISSION ON RADIATION UNITS AND MEASUREMENTS; ***"Fundamental Quantities and Units for Ionizing Radiation"**;* ICRU/Report.85ª ; Published by Oxford University Press; 2011

[27] INTERNATIONAL COMMISSION ON RADIOLOGICAL PROTECTION; ***"The 2007 Recommendation of the International Commission on Radiological Protection"***; Publication 103; Oxford; 2007

[28] NATIONAL COMMISSION ON NUCLEAR ENERGY; *"Dose Restriction, Occupational Reference Levels and Area Classification"*; Regulatory Position 3.01/004:2011; Rio de Janeiro; 2011

[29] LEVY S. D.; *"Computerisation and Unification of Radiological Protection Programmes: Monitoring Ionising Radiation"*; Thesis (PhD); IPEN - Nuclear and Energy Research Institute, São Paulo, 2015

[30] MANZOLI J. E., CAMPOS V. and DOI M.; *"Evaluation of Reproducibility and Detection Limit of CaSO4:Dy Radiation Detectors"*; Brazilian Archives of Biology and Technology; Vol 49: pp39-44; 2006

[31] INSTITUTO NACIONAL DE METROLOGIA, NORMALIZAÇÃO E QUALIDADE INDUSTRIAL; *"Evaluation of* ***measurement data - Guide to the expression of measurement uncertainty*** *- GUM 2008";* INMETRO; 1ª Brazilian Edition of the 2008 BIPM 1st Edition; Rio de Janeiro; 2012

[32] CORRÊA E.; ***"Quality Control Methodology and Implementation of Standard X-Radiation Fields, Mammography Level, Following the Iec*** *61267* ***Standard"***; master's thesis; IPEN - Nuclear and Energy Research Institute; São Paulo, 2010

[33] AMIDE: Amide's a Medical Imaging Data Examiner, Source: <http://amide.sourceforge.net> accessed on 15/04/15

[34] LOENING A.M., GAMBHIR S.S.; ***"AMIDE: A Completely Free System for Medical Imaging Data Analysis"***; Journal of Nuclear Medicine; 42(5):192P; Poster; 2001

[35] SPINKS T.J., KARIA D., LEACH M. O. and FLUX, G.; *"Quantitative PET and* ***SPECT Performance Characteristics of the AlbiraTrimodal Pre-Clinical*** *Tomography";* Physics in Medicine and Biology; 59/pag.715-731; Institute of Physics and Engineering in Medicine; 2014

[36] SÁNCHEZ F. et al; *"ALBIRA: A Small Animal PET/SPECT/CT Imaging System"*; Medical Physics; 40-051906. April 2013; American Association of Physicists in Medicine; available at: <http://dx.doi.org.ez67.periodicos.capes.gov.br/10.1118/1.4800798>; Accessed on: 04/02/2014

[37] FERNANDES H.; *"PET Image Reconstruction by Sinogram* ***Decomposition"***; master's thesis; Integrated in Biomedical Engineering; Biomedical Institute for Light and Image Research - IBILI; 2009

[38] GOERTZEN A. L. et al; *"NEMA NU 4-2008* ***Comparison of Preclinical PET*** *Imaging Systems";* Published in: Journal of Nuclear Medicine, vol. 53/n° 8, Pp. 110; 2012

ANNEX A

ipen

REGISTRO DO USO DE RADIOFÁRMACOS NO LABORATÓRIO DO MICROPET/SPEC/CT

Data	Radiofármaco	Atividade	Hora	Responsável	Obs:

Observações:

ANNEX B

ipen

REGISTRO DO USO DO EQUIPAMENTO ALBIRA MICROPET/SPCT/CT

Data	Radiofármaco	Atividade	Responsável	Tarefa		Equipamento			Estudo		Obs.:
				Hora/Inicio	Hora/Final	PET	CT	SPECT	Fantoma	Animal	

Observações:

Printed by Books on Demand GmbH, Norderstedt / Germany